T0280704

Parasitology

Volume 122 · *Supplement 2001*

Concomitant infections

EDITED BY

F. E. G. COX

CO-ORDINATING EDITOR

L. H. CHAPPELL

CAMBRIDGE
UNIVERSITY PRESS

CAMBRIDGE
UNIVERSITY PRESS

University Printing House, Cambridge CB2 8BS, United Kingdom

One Liberty Plaza, 20th Floor, New York, NY 10006, USA

477 Williamstown Road, Port Melbourne, VIC 3207, Australia

314-321, 3rd Floor, Plot 3, Splendor Forum, Jasola District Centre, New Delhi - 110025, India

79 Anson Road, #06-04/06, Singapore 079906

Cambridge University Press is part of the University of Cambridge.

It furthers the University's mission by disseminating knowledge in the pursuit of education, learning and research at the highest international levels of excellence.

www.cambridge.org
Information on this title: www.cambridge.org/9780521004978

A catalogue record for this publication is available from the British Library

ISBN 978-0-521-00497-8 Paperback

Cambridge University Press has no responsibility for the persistence or accuracy of URLs for external or third-party internet websites referred to in this publication, and does not guarantee that any content on such websites is, or will remain, accurate or appropriate.

Subscriptions may be sent to any bookseller or subscription agent or direct to the publisher: Cambridge University Press, The Edinburgh Building, Shaftesbury Road, Cambridge CB2 2RU, UK. Subscriptions in the USA, Canada and Mexico should be sent to Cambridge University Press, Journals Fulfillment Department, 110 Midland Avenue, Port Chester, NY 10573–4930, USA. All orders must be accompanied by payment. The subscription price (excluding VAT) of volumes 122 and 123, 2001 is £420 (US $696 in the USA, Canada and Mexico), payable in advance, for twelve parts plus supplements; separate parts cost £38 or US $64 each (plus postage). EU subscribers (outside the UK) who are not registered for VAT should add VAT at their country's rate. VAT registered subscribers should provide their VAT registration number. Japanese prices for institutions are available from Kinokuniya Company Ltd, P.O. Box 55, Chitose, Tokyo 156, Japan. Prices include delivery by air. Periodicals postage paid at New York, NY and at additional mailing offices. POSTMASTER: send address changes in USA, Canada and Mexico to *Parasitology*, Cambridge University Press, 110 Midland Avenue, Port Chester, New York, NY 10573–4930, USA.

Contents

List of contributions

Concomitant infections

EDITED BY F. E. G. COX

CO-ORDINATING EDITOR L. H. CHAPPELL

Preface

Concomitant infections, often referred to as mixed infections, are not only common in nature but usually more frequently encountered than infections with single organisms. Viruses, bacteria, protozoa and helminth worms are often found together in humans and wild animals yet most of what we know about host-parasite relationships are based on single organism-single host models. The collection of papers in this volume represents an attempt to summarize some of the more important aspects of host-parasite relationships as they exist in hosts harbouring one or more infectious agents with special reference to the interactions between parasites.

The papers in this volume fall into three distinct but overlapping categories, ecological, immunological and medical. Bob Poulin takes a broad-brush approach to the subject by considering the interactions that occur between species of helminths at the community level and Clive Kennedy examines, in some detail, the interactions between larval digeneans that occur within a single locality, the eye, in one host species. Frank Cox then reviews the whole field of concomitant infections from an immunological viewpoint taking representative examples from a range of experimental and natural infections and this theme is taken up and explored in more detail with special reference to nematode infections by Jerzy Behnke and his colleagues. Ian Clark then considers intraerythrocytic blood parasites and reviews the topic of heterologous immunity

and discusses the relevance of earlier reports on immunity between different species of blood parasites to our present understanding of some of the pathological processes involved in single infections. Finally in this section, Andrea Graham considers the mathematical analysis of the events that occur during concomitant infections with special reference to schistosomiasis and toxoplasmosis. The clinical aspects of concomitant infections are considered in three contributions. In the first of these, Pierre Ambroise-Thomas discusses the influence of HIV and other immunodeficiencies on parasitic infections. Lesley Drake and Don Bundy then discuss multiple infections with helminth worms in children and, in the final chapter, Peter Chiodini considers the important question faced by many clinicians as to how to treat patients with multiple parasitic infections and what the implications of such treatments might be.

The papers in this volume should be of interest to all those working with parasites whether in the field or in the laboratory and it is hoped that the information presented will be used to stimulate further research in an area that has been largely neglected. In particular, it is intended to open up the question as to whether or not the observations made under controlled laboratory conditions are relevant in the real world.

F. E. G. Cox
January 2001

Interactions between species and the structure of helminth communities

R. POULIN*

Department of Zoology, University of Otago, P.O. Box 56, Dunedin, New Zealand

SUMMARY

The role of interspecific interactions in the structure of gastrointestinal helminth communities has been at the core of most research in parasite community ecology, yet there is no consensus regarding their general importance. There have been two different approaches to the study of species interactions in helminths. The first one consists of measuring the responses of helminth species in concomitant infections, preferably in laboratory experiments. Any change in numbers of parasite individuals or in their use of niche space, compared with what is observed in single infections, provides solid evidence that the species are interacting. The second approach can only provide indirect, circumstantial evidence. It consists in contrasting observed patterns either in the distribution of species richness of infracommunities from wild hosts, in their species composition, or in pairwise associations between helminth species among infracommunities, with the random patterns predicted by appropriate null models. In many cases, observed patterns do not depart from predicted ones; when they do, alternative explanations are usually as plausible as invoking the effect of interactions among helminth species. The present evidence suggests that the role of species interactions in helminth community structure is often negligible, but that it must always be evaluated on a case-by-case basis.

Key words: Competition, exclusion, nestedness, niche segregation, species richness, saturation.

INTRODUCTION

Our knowledge of the patterns and processes underlying the structure of parasite communities has grown immensely in the past two decades (see reviews by Holmes & Price, 1986; Esch, Bush & Aho, 1990; Sousa, 1994; Poulin, 1997; Simberloff & Moore, 1997). There is still no general consensus, however, regarding the importance of interspecific interactions among parasites in the structuring of helminth communities. Some studies have suggested that helminth communities are isolationist and that the presence of one species has no influence on other species; in contrast, other studies have shown that helminth communities can be highly interactive and that species influence each other's abundance and distribution. The only attempt at a broad generalisation has been the suggestion that species interactivity is somehow linked to species richness and the average abundance of parasite individuals in hosts, or to whether the host is endothermic or not (Holmes & Price, 1986; Kennedy, Bush & Aho, 1986; Simberloff & Moore, 1997). Here I will review evidence from two parallel lines of research: experimental evidence from concomitant infections of captive hosts under laboratory conditions, and field evidence on patterns of richness and co-occurrence of parasite species from wild-caught hosts.

The study of concomitant infections intersects with parasite community ecology at the level of the individual host, i.e. at the level of the infra-

community. An infracommunity is the assemblage of all parasite individuals of all species within a single host, or within one organ in that host. The infracommunity is thus made up of all the parasite infrapopulations in a single host (*sensu* Bush *et al.* 1997). It is the interactions between infrapopulations in individual hosts that determine how many parasite species can coexist within single hosts, and ultimately in the parasite component community (all parasite individuals of all species within the host population; Bush *et al.* 1997). Much progress in parasite community ecology has been made at the level of the infracommunity, because it is usually the only level at which experimental manipulations of entire infrapopulations are possible. It is also the only level at which different parasite populations actually meet and interact in nature.

This review begins with a look at the sort of responses one can expect from parasitic helminth species interacting in concomitant infections, both on short ecological time scales and over evolutionary time. The following two sections examine the impact of species interactions on the structure of infracommunities. More precisely, these sections discuss the ways in which one can infer the existence and strength of species interactions from patterns in infracommunity structure. The bulk of the research on helminth communities has focused on gastrointestinal helminths of vertebrates; the present review is therefore restricted to these types of parasite communities. The importance of species interactions and their influence on community structure are no doubt different in other types of helminth communities, such as the communities of

* Tel: 64 3 479-7983. Fax: 64 3 479-7584.
E-mail: robert.poulin@stonebow.otago.ac.nz

Parasitology (2001), **122**, S3–S11.   Printed in the United Kingdom   © 2001 Cambridge University Press

ectoparasitic monogeneans on fish (see Rohde, 1991, 1994). Communities of larval helminths in intermediate hosts, however, do have an effect on communities of adult helminths in vertebrate definitive hosts because of the way in which helminth larvae are recruited into adult communities (Lotz, Bush & Font, 1995). Therefore, the last section of the review summarises the patterns and processes occurring in larval helminth species interacting in a shared invertebrate, intermediate host.

CONSEQUENCES OF INTERACTIONS BETWEEN HELMINTH SPECIES

There are two types of immediate consequences of species interactions that are observable in concomitant infections. These are responses occurring on short ecological time scales and best measured in experimental studies, although they can also be documented from natural infections. First, a change in the infrapopulation size of one parasite species in response to the presence of another species is a sure sign that the two species are somehow interacting, and that their numbers are not independent of one another. Second, resource use by one parasite species may change when another species is present, also an indication that they are interacting. Ecologists often give more weight to the former phenomenon, a numerical response, than to the latter one, a functional response, when assessing whether species interactions occur (Thomson, 1980). Generally, in studies of helminth parasites, either or both numerical and functional responses are taken as evidence of interaction (Poulin, 1998). The possible scenarios are illustrated in Fig. 1.

Numerical responses are not only quite common in mixed infections of gastrointestinal helminths, but also often very substantial, with infrapopulations of one species reduced by as much as half the size they achieve when not sharing the host with another species (compare Fig. 1A and B; see examples in Dobson, 1985; Poulin, 1998). Note that numerical responses need not be negative; however, competitive interactions appear more common among helminths than positive interactions. When they are extremely strong, competitive interactions can lead to the exclusion of one species by the other, the most extreme numerical effect possible. An interesting feature of these numerical responses is their asymmetrical nature. Typically, one helminth species incurs severe reductions in numbers whereas the other is almost unaffected (e.g. Dash, 1981; Holland, 1984). A similar pattern emerges when it is reductions in average worm fecundity rather than reductions in infrapopulation sizes that are the numerical responses measured (e.g. Silver, Dick & Welch, 1980; Holland, 1984). Thus there is often a winner and a loser, an outcome that has at least two possible explanations. First, the interactions may be

closer to one-sided interference competition than classical exploitative competition, with one species intrinsically favoured when interacting with the other one. Second, there may be a priority effect, with the outcome of the interaction being dependent on which species first becomes established in the host; in this situation, neither species is intrinsically a winner, and all depends on which species gets a head start.

These types of large numerical responses are frequently observed in experimental infections, but the significance of the interspecific interactions uncovered in the laboratory for the structuring of helminth communities in nature may be much lower. The typical distribution of helminth numbers among host individuals follows an aggregated pattern (Shaw & Dobson, 1995), in which most hosts harbour few or no parasites and only few hosts harbour large infrapopulations. Assuming that the aggregated distributions of different parasite species are independent of one another, there may be few opportunities for two or more species of parasites to co-occur in the same host individual in sufficient numbers for competition and numerical responses to take place. Indeed, mathematical models (Dobson, 1985; Dobson & Roberts, 1994) and empirical studies (Morand et al. 1999) indicate that parasite aggregation may dampen the effects of interspecific competition. The possibility that parasites alter their use of resources when co-occurring with potential competitors also renders numerical responses less likely.

Like numerical responses, functional responses are best studied experimentally, although they can be inferred from natural infections. The evidence of antagonistic interactions provided by functional responses is not as solid as that from numerical responses (Thomson, 1980), but it can be very suggestive. In studies of gastrointestinal helminth parasites, the most widely studied type of functional response is a shift in the site of attachment of helminths in the host gut. This is viewed as a simplified measure of the niche of the parasite, focusing only on the spatial dimension of the niche because it is simpler to quantify. It then becomes possible to define the fundamental niche of a parasite species as the precise region of the gut it inhabits when in single infections, i.e. when not sharing the host with other parasite species. The fundamental niche represents the preferred range of infection sites among those where the parasite can develop successfully. When the parasite co-occurs with a competing species, it may alter its distribution in the host gut in order to minimise its spatial overlap with the competitor. The niche it then occupies is called its realised niche; it is often the portion of the fundamental niche that is competitor-free. The shift from the fundamental to the realised niche is the basic functional response to competition (compare

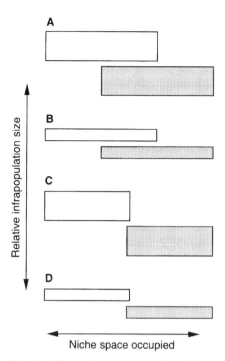

Fig. 1. Schematic representation of the possible effects of concomitant infections on the infrapopulations of two parasite species. Hypothetical infection doses, i.e. the number of parasite larvae of each species administered to hosts, are kept constant in all cases. The vertical dimension of the rectangles illustrates the relative size of each infrapopulation, while the horizontal dimension indicates the use of niche space by each infrapopulation. A, the infrapopulation sizes and fundamental niches of the two species when in single infections; B, a numerical response to concomitant infection, in which infrapopulation sizes are reduced; C, a functional response to concomitant infection, in which the realised niches of both parasites are adjusted to reduce niche overlap; and D, a joint numerical and functional response.

Fig. 1A and C), which Holmes (1973) called interactive site segregation. It has been beautifully illustrated in classical experimental studies of single-species and mixed-species infections (Holmes, 1961; Patrick, 1991). Functional responses can also be investigated by studying the niches of intestinal helminths in naturally infected hosts (e.g. Bush & Holmes, 1986b; Haukisalmi & Henttonen, 1993b; Ellis, Pung & Richardson, 1999). Competing parasite species may display either numerical or functional responses as a consequence of their interaction, or both responses simultaneously (compare Fig. 1A and D).

The realised niche of a parasite species co-occurring with competing species may not coincide with its preferred or optimal part of the fundamental niche. If the nutrient supply or other conditions are of lower quality in the realised niche than in the preferred portion of the fundamental niche, there will be a cost associated with the functional response. For the response to be favoured, this cost must be

lower than that associated with staying in place and competing with other parasite species. Evidence from species-rich intestinal helminth communities of birds suggests that a functional response is usually the best strategy, since realised niches of helminths are almost always more restricted than their fundamental niches (Bush & Holmes, 1986b; Stock & Holmes, 1988). However, other studies on gastro-intestinal helminths with overlapping fundamental niches have not systematically found functional responses in competitive situations, whether or not numerical responses were observed (Moore & Simberloff, 1990; Ellis et al. 1999).

Finally, it is also possible that co-occurring species do not interact, either because they are not abundant enough to exert mutual pressures on one another, or because they differ in resource use and their fundamental niches do not overlap. There are many reasons for such differences in realised niches between species, many of which have nothing to do with species interactions. For instance, the reproductive success of helminths may vary as a function of each worm's position in the gut (Sukhdeo, 1991), and selection may have favoured a narrowing of the niche around sites where fitness is maximised. This could produce isolationist parasite communities, in which functional responses do not occur and the breadth and location of the niche of one species is independent of the presence of other species. However, the difference in resource use can be the product of an evolutionary niche shift that was selected as a result of intense competition in the past (Holmes, 1973). In other words, intense competition between common and abundant species will favour the evolutionary divergence of their fundamental niches, leaving only the ghost of competition past (sensu Connell, 1980). In all such scenarios, present-day interactions between helminth species would not be detectable and would play no role in structuring the parasite community.

SPECIES RICHNESS IN HELMINTH INFRACOMMUNITIES

If there are interactions between helminth species, we should expect that they would impact on the number of species in helminth communities and on their distribution among host individuals. The effect of interactions should be to push the observed patterns away from those expected from random, non-interactive assembly rules. In this section, I will review how the existence of species interactions can be inferred from patterns in the richness of infra-communities in natural systems; in the following section, I will discuss what the observed patterns in species composition and associations in infra-communities can tell us about the action of species interactions.

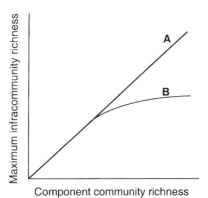

Fig. 2. Two possible relationships between the maximum (or mean) infracommunity species richness and the richness of the component community, across host populations or related host species. A, a linear relationship indicates that infracommunity richness increases proportionately with the richness of the component community, possibly because infracommunities are non-interactive; B, a saturation curve indicates that infracommunity richness reaches a limit and becomes independent of component community richness, perhaps as a result of species saturation and strong interspecific interactions.

Individual hosts from the same population do not all harbour the same number of parasite species. Some hosts may harbour most of the parasite species included in the component parasite community, others only one or two species. Characteristics of these hosts, such as age, size or preferred foraging site, may account for the variation among them in helminth species richness, by influencing their likelihood of acquiring certain parasite species. Among a random sample of hosts, however, and if hosts are random samplers of the parasites available in their habitat, one would generally expect the frequency distribution of parasite species richness values among hosts (i.e. among infracommunities) to follow a null model if species interactions are not important. If mechanisms such as competitive exclusion are acting, though, one would expect the distribution of richness values to depart from that expected by chance. Null models can be constructed for any particular system according to the total number of parasite species in the component community and their respective prevalences (Janovy et al. 1995). Comparisons of observed frequency distributions of helminth species richness among vertebrate hosts with those predicted by a null model indicate that the null model accounts for observed data in well over half of published studies (Poulin, 1996, 1997). Thus in the majority of cases, there is no evidence of interactions between helminth species in natural assemblages. Where departures from the null expectations do occur, species interactions are only one of many possible causes.

Recently, Dove (1999) proposed an index of interactivity based on the non-linear relationship between the number of hosts examined and the estimated component community richness. In a sample of hosts, the number of recorded parasite species increases with the number of hosts examined until it reaches an asymptote corresponding to the true component community richness. Fitting a growth curve to observed data creates two key parameters, the asymptote and the gradient of the curve before the asymptote is reached. This gradient is related to the mean infracommunity richness: as the mean infracommunity richness increases and approaches the actual component community richness, the gradient of the slope becomes steeper. Dove (1999) suggests that the product of the two terms provides a measure of the degree of interactivity in a parasite community. High values of this index would indeed characterize communities traditionally viewed as interactive, i.e. those with many species with high prevalences (Holmes & Price, 1986). This provides another example of how patterns in infracommunity richness can be used to assess the role of species interactions in structuring parasite communities. Dove's (1999) index, however, will require some testing before its wide acceptance; it may reflect species saturation in infracommunities more than species interactions.

An investigation of species saturation itself can reveal the existence of species interactions if these are strong enough to lead to competitive species exclusion. The relationship between infracommunity richness and component community richness can be viewed as that between local and regional species richness in free-living communities (Cornell & Lawton, 1992; Srivastava, 1999). Two extreme scenarios are possible (Fig. 2). First, the relationship may be linear, suggesting that the number of species in an infracommunity is generally proportional to the number available in the component community, and that species interactions are negligible. Second, a curvilinear relationship would indicate that infracommunity richness becomes increasingly independent of component community richness as the latter increases, a phenomenon that may be due to interspecific interactions and the species saturation of infracommunities. The maximum number of species that can occur in an infracommunity is equal to the number of species in the component community. Typically, in species-rich component communities, no single infracommunity includes all available parasite species. On its own, however, a curvilinear relationship between local (infracommunity) and regional (component community) species richness is not sufficient to conclude that interspecific interactions lead to species saturation (Srivastava, 1999). Other lines of evidence, such as an experimental demonstration of competitive effects, are required to support any pattern in local and regional species richness.

This approach has been applied to parasite communities recently. In comparative studies of gastrointestinal communities in different species of birds and mammals, the relationship between maximum infracommunity richness and component community richness came out as clearly linear (Poulin, 1998). In a study of 64 component communities of intestinal helminths from the same host fish species, however, the relationship proved to be curvilinear (Kennedy & Guégan, 1996). Here infracommunity saturation is a possible explanation because previous studies have demonstrated the existence of competitive interactions between pairs of parasite species (Kennedy & Guégan, 1996). Rohde (1998) suggests that this and other curvilinear relationships can also result from processes that have nothing to do with either interspecific interactions or saturation, and that no strong conclusions can be drawn from these results until null expectations can be generated that will take other processes into account. Rohde's (1998) views are based on computer simulations that are maybe a bit simplistic; nevertheless, caution is required in the interpretation of saturation curves.

Inferring processes, such as interspecific interactions between helminth species, from patterns in community parameters is not without difficulties. The attempts at doing so have not produced any compelling evidence for a generally important role for these interactions in determining infracommunity richness in natural assemblages. The next section will summarise attempts at inferring the action of species interactions from the composition and structure of infracommunities.

STRUCTURE OF HELMINTH INFRACOMMUNITIES

When the helminth species forming each infracommunity are random subsets of the ones available in the component community, the obvious conclusion is that there are no forces acting to structure infracommunities. However, evidence that the composition of infracommunities is not random would suggest that structuring processes, one of which may be interactions among helminth species, are at work. As in the previous section, the main approach used to detect species interactions or other structuring forces consist of comparisons between observed patterns in natural assemblages and those expected from null models of species associations. Null models are useful because they provide a baseline for such comparisons; however, they must be chosen with great care so that they generate realistic and appropriate null patterns (Gotelli & Graves, 1996).

Nestedness in species composition is a common departure from randomness in free-living assemblages occupying insular or fragmented habitats (Patterson & Atmar, 1986; Worthen, 1996; Wright *et al.* 1998). In a parasite component community, a nested pattern would mean that the species forming a species-poor infracommunity are distinct subsets

of progressively richer infracommunities (Fig. 3). In other words, parasite species with high prevalences would be found in all sorts of infracommunities, whereas rare parasite species would only occur in species-rich infracommunities. Nestedness is a departure from a random assembly of species, and is tested against the expectations derived from a null model based on each species' prevalence (Patterson & Atmar, 1986). Of course, all kinds of structuring processes can generate nested patterns; in the context of interspecific interactions, finding a nested pattern would only indicate that structuring of the infracommunities is taking place, nothing more. Nestedness has been investigated in communities of fish ectoparasites (Guégan & Hugueny, 1994; Worthen & Rohde, 1996), but the only nestedness analyses of gastrointestinal helminth communities of vertebrates were done on 2 mammal species and 8 marine fish species (Poulin, 1996; Rohde *et al.* 1998). Of the 10 component communities tested, significant nestedness was only observed in 2 communities of marine fish. Similarly, nestedness does not appear to be common among communities of metazoan ectoparasites of fish (Worthen & Rohde, 1996; Rohde *et al.* 1998). Thus to date these analyses tend to indicate that the composition of infracommunities does not usually follow a nested pattern, which could mean that the actions of interspecific interactions or other structuring processes are not apparent. However, a considerable number of the parasite communities investigated by Rohde *et al.* (1998) suggest (based on the P values and how they were derived) that another non-random pattern is common, one that may be best referred to as anti-nestedness (R. Poulin & J.-F. Guégan in press); this strange pattern will require further study, but it may yet serve to highlight the existence of interactions or other structuring forces.

The associations between different pairs of helminth species among infracommunities is by far the aspect of parasite community structure that has received the most attention to date (e.g. Andersen & Valtonen, 1990; Moore & Simberloff, 1990; Lotz & Font, 1991, 1994; Haukisalmi & Henttonen, 1993*a*). Positive and negative associations between helminth species, that is co-occurrences of pairs of species that are more or less frequent than expected by chance, respectively, can provide strong evidence that species interactions exist and act on community structure. These associations can be computed using presence/absence data, or preferably using actual parasite numbers or biomass. A common assumption (the null model) in tests of pairwise associations among species is that the number of positive covariances should equal the number of negative ones if infracommunities are assembled randomly. In many systems, the number of positive associations among helminths in infracommunities exceeds the number of negative ones, suggesting that positive interactions

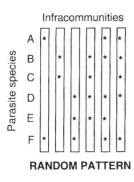

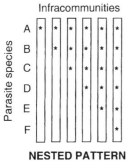

Fig. 3. Two hypothetical distributions of parasite species among infracommunities (i.e. among host individuals) in a component community. Each rectangle represents a different infracommunity, and infracommunities are arranged from least (at left) to most (at right) species-rich. The average infracommunity richness and the average prevalence of the six parasite species are the same in the two examples of presence/absence matrices. In a perfectly nested design, a parasite species occurring in a host individual with n parasite species will be found in all host individuals with at least $n+1$ species (adapted from Poulin, 1998).

among species structure the community (e.g. Bush & Holmes, 1986a; Lotz & Font, 1991). However, several factors can generate spurious covariances and bias the sign of associations, and a precise null model must be used to assess whether positive or negative interactions are predominant. Lotz & Font (1994) showed that a high proportion of rare parasite species, with low prevalence in the component community, can produce an excess of spurious negative associations, whereas a high proportion of common species with high prevalence can lead to an excess of positive associations. In addition, they showed that the number of hosts (infracommunities) sampled also influences the likelihood of obtaining false pairwise associations between species. Using randomisation procedures to generate a more precise null model, Lotz & Font (1994) showed that the excess in positive associations that they had observed in an earlier study of helminths parasitic in bats (Lotz & Font, 1991), were unlikely to reflect a predominance of positive interactions between species.

Even if positive or negative associations are more numerous that predicted by a refined null model, they may have several explanations. First, they may

indeed reflect the action of interspecific interactions on community structure. Negative associations could result from strong interspecific competition which can lead to the exclusion of one species by another; positive associations could result from facilitation processes, such as immunosuppression induced by one species and benefiting other species. Second, most statistical methods used to detect species covariances are more sensitive to positive associations that to negative ones, which can bias any analysis of pairwise associations (Haukisalmi & Henttonen, 1998). Third, heterogeneity among host individuals can lead to certain hosts acquiring certain sets of parasite species and other hosts acquiring others, without species interactions among parasites being necessary. Fourth, similarity between helminth species with respect to the time of year at which they are acquired by vertebrates, combined with long durations of infection, can generate positive associations between species (Forbes et al. 1999).

One other process can generate patterns of species associations among species of gastrointestinal helminths in vertebrate hosts. With rare exceptions, helminths join an infracommunity when their larvae are ingested by the host. Larvae of acanthocephalans, cestodes, digeneans and many nematodes are acquired when the definitive host preys on an intermediate host which may contain larvae of more than one helminth species. Thus larvae arrive in packets, not singly (Bush, Heard & Overstreet, 1993). The recruitment of one species may not be independent of other species, if larvae are positively or negatively associated in intermediate hosts. The structure of larval helminth communities can be transferred to adult helminth communities in vertebrates. Using computer simulations, Lotz et al. (1995) found that pairwise associations in intermediate hosts can be transferred to definitive hosts, especially positive ones. Therefore observed species associations among gastrointestinal helminths in definitive hosts may have nothing to do with species interactions or other processes operating in definitive hosts. They may, however, reflect interactions among larval helminths in intermediate hosts. It is thus time to look briefly at what is known of species interactions among larval helminths in their intermediate, invertebrate hosts.

HELMINTH COMMUNITIES IN INTERMEDIATE HOSTS

There are basic differences between larval helminth communities in intermediate hosts and adult helminth communities in definitive hosts. First, communities in intermediate hosts usually comprise much fewer species than those in definitive hosts. Second, the prevalences of helminth species in their intermediate hosts are typically very low (often

< 1 %; see review in Marcogliese, 1995). Third, many helminth larval stages, such as the metacercariae of digeneans (but not the sporocysts or rediae), the cystacanths of acanthocephalans, or the cysticercoids of cestodes, consume few host resources and are mostly inactive. These differences might suggest that competition or other types of species interactions are unlikely to be important in invertebrate intermediate hosts.

In fact, the available evidence suggests otherwise. There have been few experimental studies of species interactions between helminths in intermediate hosts. However, these few studies show strong effects of larvae of one species on the establishment or survival of larvae of a second species (Gordon & Whitfield, 1985; Barger & Nickol, 1999). Field studies examining patterns of mixed infections in wild intermediate hosts have all focused on larval digeneans in their gastropod intermediate hosts. In a comprehensive review of field surveys, Kuris & Lafferty (1994) found that strong asymmetrical numerical effects were the norm, with dominant larval digenean species typically causing substantial reductions in the abundance of subordinate species, if not causing their complete exclusion. They found that double or mixed infections of snails were much less frequent than expected based on the respective prevalences of digenean species in a community. Competition, and even predation, are common interspecific interactions within infracommunities of larval digeneans in their molluscan hosts (Sousa, 1992, 1993). It must be pointed out, though, that interspecific interactions between larval digeneans may be unimportant in systems where different species only co-occur infrequently (Curtis & Hubbard, 1993; Curtis, 1997), or in systems where infection of snails by the different parasite species is subject to extensive spatial or temporal heterogeneity (Esch *et al.* 1997).

We may also expect strong positive associations between larval helminths in systems where one larval helminth species is capable of manipulating the phenotype of the intermediate host to increase its probability of being ingested by a suitable definitive host. Host manipulation by parasites is a relatively common phenomenon that can greatly improve transmission rates of helminths with complex life cycles (Poulin, 1998). Parasite species sharing an intermediate host population with a manipulator species would benefit by associating with the manipulator, as they would obtain a cost-free ride to the definitive host, assuming they also share the latter with the manipulator (Thomas, Renaud & Poulin, 1998; Lafferty, 1999). The few tests of this idea have generally found the predicted positive association between the manipulator and the 'hitch-hiker' (Lafferty, Thomas & Poulin, 2000). Similarly, negative associations may be expected between larval parasites with conflicting interests, such as two

species capable of manipulating intermediate hosts but with different definitive host species. In these cases, larvae of one species would benefit by avoiding larvae of the other species because their destinations are different (Lafferty *et al.* 2000). Thus both negative and positive associations, resulting or not from species interactions, can occur in intermediate hosts and be transferred to definitive hosts, rendering more difficult the detection of interactions between adult helminths from non-random patterns in infracommunity structure.

CONCLUSIONS

In experimental studies, the importance of species interactions in concomitant infections can be demonstrated clearly in a straightforward manner: when two species co-occur, either there are numerical or functional effects, or there are not. Since effects have often been observed, it is natural to ask whether species interactions play a role in structuring natural parasite communities. This question lies on a grander scale than that of a two-species laboratory system, and cannot easily be addressed experimentally. Studies of natural patterns in parasite communities have been slowly accumulating, mainly following the application of concepts (e.g. the relationship between local and regional species richness, and nestedness) initially developed for communities of free-living organisms. When trying to infer the existence and importance of species interactions from patterns in the structure of natural parasite communities, however, two ubiquitous problems make all conclusions tentative. First, observed patterns must depart from randomness to be meaningful, and it has not always been possible to devise appropriate null models for comparisons with observed patterns. Second, several explanations other than the effect of species interactions can account for the discrepancies between observed and null patterns, making it impossible to determine whether there are any interactions. To be rigorous, one should always obtain a direct demonstration of species interactions to conclude that they exist and influence community structure. In other words, to use a metaphor, departures from random patterns of community structure are only circumstantial evidence, a controlled experiment is an eyewitness account, and an observed response during concomitant infection is a smoking gun.

Sousa (1994) reviewed quantitative studies of helminth infracommunity structure to see if they supported the general notion that interactions would be important in the dense, speciose infracommunities of endothermic vertebrates (Holmes & Price, 1986; Kennedy *et al.* 1986). He found that while some systems fit the predictions rather well, it was not possible to accurately predict whether a community would be interactive or isolationist based on its species richness or the nature of its host. Broad generalisations are still impossible at this time, and a case-by-case approach remains the best way to assess the importance of species interactions in helminth communities.

ACKNOWLEDGEMENTS

I am grateful to J.-F. Guégan, C. R. Kennedy, S. Morand, and E. T. Valtonen for useful comments on an earlier draft.

REFERENCES

ANDERSEN, K. I. & VALTONEN, E. T. (1990). On the infracommunity structure of adult cestodes in freshwater fishes. *Parasitology* **101**, 257–264.

BARGER, M. A. & NICKOL, B. B. (1999). Effects of coinfection with *Pomphorhynchus bulbocolli* on development of *Leptorhynchoides thecatus* (Acanthocephala) in amphipods (*Hyalella azteca*). *Journal of Parasitology* **85**, 60–63.

BUSH, A. O., HEARD, R. W. & OVERSTREET, R. M. (1993). Intermediate hosts as source communities. *Canadian Journal of Zoology* **71**, 1358–1363.

BUSH, A. O. & HOLMES, J. C. (1986a). Intestinal helminths of lesser scaup ducks: patterns of association. *Canadian Journal of Zoology* **64**, 132–141.

BUSH, A. O. & HOLMES, J. C. (1986b). Intestinal helminths of lesser scaup ducks: an interactive community. *Canadian Journal of Zoology* **64**, 142–152.

BUSH, A. O., LAFFERTY, K. D., LOTZ, J. M. & SHOSTAK, A. W. (1997). Parasitology meets ecology on its own terms: Margolis *et al.* revisited. *Journal of Parasitology* **83**, 575–583.

CONNELL, J. H. (1980). Diversity and the coevolution of competitors, or the ghost of competition past. *Oikos* **35**, 131–138.

CORNELL, H. V. & LAWTON, J. H. (1992). Species interactions, local and regional processes, and limits to the richness of ecological communities: a theoretical perspective. *Journal of Animal Ecology* **61**, 1–12.

CURTIS, L. A. (1997). *Ilyanassa obsoleta* (Gastropoda) as a host for trematodes in Delaware estuaries. *Journal of Parasitology* **83**, 793–803.

CURTIS, L. A. & HUBBARD, K. M. K. (1993). Species relationships in a marine gastropod-trematode ecological system. *Biological Bulletin* **184**, 25–35.

DASH, K. M. (1981). Interaction between *Oesophagostomum columbianum* and *Oesophagostomum venulosum* in sheep. *International Journal for Parasitology* **11**, 201–207.

DOBSON, A. P. (1985). The population dynamics of competition between parasites. *Parasitology* **92**, 675–682.

DOBSON, A. P. & ROBERTS, M. (1994). The population dynamics of parasitic helminth communities. *Parasitology* **109**, S97–S108.

DOVE, A. D. M. (1999). A new index of interactivity in parasite communities. *International Journal for Parasitology* **29**, 915–920.

ELLIS, R. D., PUNG, O. J. & RICHARDSON, D. J. (1999). Site selection by intestinal helminths of the Virginia opossum (*Didelphis virginiana*). *Journal of Parasitology* **85**, 1–5.

ESCH, G. W., BUSH, A. O. & AHO, J. M. (1990). *Parasite Communities: Patterns and Processes.* London, Chapman & Hall.

ESCH, G. W., WETZEL, E. J., ZELMER, D. A. & SCHOTTHOEFER, A. M. (1997). Long-term changes in parasite population and community structure: a case history. *American Midland Naturalist* **137**, 369–387.

FORBES, M. R., ALISAUSKAS, R. T., MCLAUGHLIN, J. D. & CUDDINGTON, K. M. (1999). Explaining co-occurrence among helminth species of lesser snow geese (*Chen caerulescens*) during their winter and spring migration. *Oecologia* **120**, 613–620.

GORDON, D. M. & WHITFIELD, P. J. (1985). Interactions of the cysticercoids of *Hymenolepis diminuta* and *Raillietina cesticillus* in their intermediate host, *Tribolium confusum*. *Parasitology* **90**, 421–431.

GOTELLI, N. J. & GRAVES, G. R. (1996). *Null Models in Ecology*. Washington, D.C., Smithsonian Institution Press.

GUÉGAN, J.-F. & HUGUENY, B. (1994). A nested parasite species subset pattern in tropical fish: host as major determinant of parasite infracommunity structure. *Oecologia* **100**, 184–189.

HAUKISALMI, V. & HENTTONEN, H. (1993a). Coexistence in helminths of the bank vole *Clethrionomys glareolus*. I. Patterns of co-occurrence. *Journal of Animal Ecology* **62**, 221–229.

HAUKISALMI, V. & HENTTONEN, H. (1993b). Coexistence in helminths of the bank vole *Clethrionomys glareolus*. II. Intestinal distribution and interspecific interactions. *Journal of Animal Ecology* **62**, 230–238.

HAUKISALMI, V. & HENTTONEN, H. (1998). Analysing interspecific associations in parasites: alternative methods and effects of sampling heterogeneity. *Oecologia* **116**, 565–574.

HOLLAND, C. (1984). Interactions between *Moniliformis* (Acanthocephala) and *Nippostrongylus* (Nematoda) in the small intestine of laboratory rats. *Parasitology* **88**, 303–315.

HOLMES, J. C. (1961). Effects of concurrent infections on *Hymenolepis diminuta* (Cestoda) and *Moniliformis dubius* (Acanthocephala). I. General effects and comparison with crowding. *Journal of Parasitology* **47**, 209–216.

HOLMES, J. C. (1973). Site segregation by parasitic helminths: interspecific interactions, site segregation, and their importance to the development of helminth communities. *Canadian Journal of Zoology* **51**, 333–347.

HOLMES, J. C. & PRICE, P. W. (1986). Communities of parasites. In *Community Ecology: Pattern and Process* (ed. Anderson, D. J. & Kikkawa, J.), pp. 187–213. Oxford, Blackwell Scientific Publications.

JANOVY, J. Jr., CLOPTON, R. E., CLOPTON, D. A., SNYDER, S. D., EFTING, A. & KREBS, L. (1995). Species density distributions as null models for ecologically significant interactions of parasite species in an assemblage. *Ecological Modelling* **77**, 189–196.

KENNEDY, C. R., BUSH, A. O. & AHO, J. M. (1986). Patterns in helminth communities: why are birds and fish so different? *Parasitology* **93**, 205–215.

KENNEDY, C. R. & GUÉGAN, J.-F. (1996). The number of niches in intestinal helminth communities of *Anguilla*

anguilla: are there enough spaces for parasites? *Parasitology* **113**, 293–302.

KURIS, A. M. & LAFFERTY, K. D. (1994). Community structure: larval trematodes in snail hosts. *Annual Review of Ecology and Systematics* **25**, 189–217.

LAFFERTY, K. D. (1999). The evolution of trophic transmission. *Parasitology Today* **15**, 111–115.

LAFFERTY, K. D., THOMAS, F. & POULIN, R. (2000). Evolution of host phenotype manipulation by parasites and its consequences. In *Evolutionary Biology of Host-Parasite Relationships: Theory Meets Reality* (ed. Poulin, R., Morand, S. & Skorping, A.), pp. 117–127. Amsterdam, Elsevier Science Publishers.

LOTZ, J. M., BUSH, A. O. & FONT, W. F. (1995). Recruitment-driven, spatially discontinuous communities: a null model for transferred patterns in target communities of intestinal helminths. *Journal of Parasitology* **81**, 12–24.

LOTZ, J. M. & FONT, W. F. (1991). The role of positive and negative interspecific associations in the organization of communities of intestinal helminths of bats. *Parasitology* **103**, 127–138.

LOTZ, J. M. & FONT, W. F. (1994). Excess positive associations in communities of intestinal helminths of bats: a refined null hypothesis and a test of the facilitation hypothesis. *Journal of Parasitology* **80**, 398–413.

MARCOGLIESE, D. J. (1995). The role of zooplankton in the transmission of helminth parasites to fish. *Reviews in Fish Biology and Fisheries* **5**, 336–371.

MOORE, J. & SIMBERLOFF, D. (1990). Gastrointestinal helminth communities of bobwhite quail. *Ecology* **71**, 344–359.

MORAND, S., POULIN, R., ROHDE, K. & HAYWARD, C. (1999). Aggregation and species coexistence of ectoparasites of marine fishes. *International Journal for Parasitology* **29**, 663–672.

PATRICK, M. J. (1991). Distribution of enteric helminths in *Glaucomys volans* L. (Sciuridae): a test for competition. *Ecology* **72**, 755–758.

PATTERSON, B. D. & ATMAR, W. (1986). Nested subsets and the structure of insular mammalian faunas and archipelagos. *Biological Journal of the Linnean Society* **28**, 65–82.

POULIN, R. (1996). Richness, nestedness, and randomness in parasite infracommunity structure. *Oecologia* **105**, 545–551.

POULIN, R. (1997). Species richness of parasite assemblages: evolution and patterns. *Annual Review of Ecology and Systematics* **28**, 341–358.

POULIN, R. (1998). *Evolutionary Ecology of Parasites: From Individuals to Communities.* London, Chapman & Hall.

POULIN, R., GUÉGAN, J.-F. (in press). Nestedness, anti-nestedness, and the relationship between prevalence and intensity in ectoparasite assemblages of marine fish: a spatial model of species coexistence. *International Journal for Parasitology.*

ROHDE, K. (1991). Intra- and interspecific interactions in low density populations in resource-rich habitats. *Oikos* **60**, 91–104.

ROHDE, K. (1994). Niche restriction in parasites: proximate and ultimate causes. *Parasitology* **109**, S69–S84.

ROHDE, K. (1998). Is there a fixed number of niches for endoparasites of fish? *International Journal for Parasitology* **28**, 1861–1865.

ROHDE, K., WORTHEN, W. B., HEAP, M., HUGUENY, B. & GUÉGAN, J.-F. (1998). Nestedness in assemblages of metazoan ecto- and endoparasites of marine fish. *International Journal for Parasitology* **28**, 543–549.

SHAW, D. J. & DOBSON, A. P. (1995). Patterns of macroparasite abundance and aggregation in wildlife populations: a quantitative review. *Parasitology* **111**, S111–S133.

SILVER, B. B., DICK, T. A. & WELCH, H. E. (1980). Concurrent infections of *Hymenolepis diminuta* and *Trichinella spiralis* in the rat intestine. *Journal of Parasitology* **66**, 786–791.

SIMBERLOFF, D. & MOORE, J. (1997). Community ecology of parasites and free-living animals. In *Host-Parasite Evolution: General Principles and Avian Models* (ed. Clayton, D. H. & Moore, J.), pp. 174–197. Oxford, Oxford University Press.

SOUSA, W. P. (1992). Interspecific interactions among larval trematode parasites of freshwater and marine snails. *American Zoologist* **32**, 583–592.

SOUSA, W. P. (1993). Interspecific antagonism and species coexistence in a diverse guild of larval trematode parasites. *Ecological Monographs* **63**, 103–128.

SOUSA, W. P. (1994). Patterns and processes in communities of helminth parasites. *Trends in Ecology & Evolution* **9**, 52–57.

SRIVASTAVA, D. S. (1999). Using local-regional richness plots to test for species saturation: pitfalls and potentials. *Journal of Animal Ecology* **68**, 1–16.

STOCK, T. M. & HOLMES, J. C. (1988). Functional relationships and microhabitat distributions of enteric helminths of grebes (Podicipedidae): the evidence for interactive communities. *Journal of Parasitology* **74**, 214–227.

SUKHDEO, M. V. K. (1991). The relationship between intestinal location and fecundity in adult *Trichinella spiralis*. *International Journal for Parasitology* **21**, 855–858.

THOMAS, F., RENAUD, F. & POULIN, R. (1998). Exploitation of manipulators: 'hitch-hiking' as a parasite transmission strategy. *Animal Behaviour* **56**, 199–206.

THOMSON, J. D. (1980). Implications of different sorts of evidence for competition. *American Naturalist* **116**, 719–726.

WORTHEN, W. B. (1996). Community composition and nested-subset analyses: basic descriptors for community ecology. *Oikos* **76**, 417–426.

WORTHEN, W. B. & ROHDE, K. (1996). Nested subset analyses of colonization-dominated communities: metazoan ectoparasites of marine fishes. *Oikos* **75**, 471–478.

WRIGHT, D. H., PATTERSON, B. D., MIKKELSON, G. M., CUTLER, A. & ATMAR, W. (1998). A comparative analysis of nested subset patterns of species composition. *Oecologia* **113**, 1–20.

Interspecific interactions between larval digeneans in the eyes of perch, *Perca fluviatilis*

C. R. KENNEDY*

School of Biological Sciences, University of Exeter, Exeter EX4 4PS, UK

SUMMARY

The changes in prevalence and abundance of the three species of metacercariae in the humour of the eyes of perch *Perca fluviatilis* in Slapton Ley, Devon, have been monitored over a period of 29 years. Earlier studies had revealed that *Diplostomum gasterostei* was originally the sole occupant of this niche, but *Tylodelphys clavata* colonised in 1973 and *T. podicipina* in 1976. A decline in the number of perch with heavy infections of *D. gasterostei* was significantly negatively correlated with population abundance of *T. clavata* and a decline in recruitment rate of *D. gasterostei* coincided with the population increase in *T. podicipina* over the period 1976–1979. It was suggested that the decline in population size of *D. gasterostei* was due to inter-specific competition, but this hypothesis could not be tested experimentally. Subsequent investigations, reported here, confirmed the decline when the data set was extended to 1985. A severe decline in the perch population over the winter of 1984/1985 resulted in the disappearance of *D. gasterostei* and *T. podicipina* and this was followed by a slow recovery from 1990 onwards. This natural experiment provided an opportunity to test the hypothesis. Only *T. clavata* survived throughout the perch crash and the population continued at pre-crash levels up to 1999. Its congener *T. podicipina* did not re-appear until 1994 and was probably a re-introduction: it did not attain pre-crash levels until 1999. It is likely that *D. gasterostei* survived the crash as it re-appeared in 1991, but was confined to young of the year fish and barely approached pre-crash levels even in 1999. Its continual low levels cannot be explained by changes in the lake or in densities of snail intermediate or bird definitive hosts. New data revealed that the suspensory ligaments of the eye were the preferred site of all three species and that the eye was partitioned out between them. The data from the post-crash period do not refute but rather confirm earlier conclusions that inter-specific competition is responsible for the decline in *D. gasterostei* and this remains the preferred hypothesis.

Key words: Metacercariae, interspecific competition, population dynamics, niche, resource partitioning, negative interactions.

INTRODUCTION

Interspecific interactions and their role in structuring helminth communities in freshwater fish are currently a topical issue (Bush & Holmes, 1986; Stock & Holmes, 1987, 1988; Rohde, 1979, 1991, 1998; Kennedy & Guégan, 1996; Kennedy & Hartvigsen, 2000; Poulin, this volume). There is evidence from mammalian hosts that interspecific interactions may involve immune responses (Holland, 1984, 1987; see also Behnke *et al.* and Cox, this volume) in one way or another, but there is little evidence that these are of importance in fish hosts. Evidence for interspecific interactions between helminths in fish is seldom direct. Interactions have been reported for monogeneans on gills and generally are deduced from observations on the decline of one species in the presence of another (Kennedy & DiCave, 1998) or from a niche shift or resource partitioning between species within a habitat (Buchmann, 1988). In the view of some workers, notably Rohde (1979, 1991, 1998), they are of infrequent occurrence and little or no ecological significance as niche restriction can be explained in other ways.

Inter-specific interactions between species in fish intestines have often been deduced from evidence of niche shifts by one species in the presence of another (Chappell, 1969; Halvorsen & Macdonald, 1972; Grey & Hayunga, 1980) or, more rarely, from the decline of one species in the presence of another (Kennedy, 1992) or the exclusion of one species by another (Kennedy & Hartvigsen, 2000). Direct, experimental evidence for interspecific interactions is difficult to obtain and hence very scarce, the studies of Bates & Kennedy (1990, 1991) being a rare exception. Most of the known examples involve an acanthocephalan and/or a cestode and in no case has the resource for which the species compete been identified with certainty, nor is the mechanism of the competition understood. It is frequently suggested that there is direct competition for nutrients or for intestinal space which, in turn, may provide access to nutrients. It is generally accepted that interspecific competition is not common, partly because resource partitioning in space is common and will serve to avoid it (Holmes, 1973) and partly because co-existence of potential competitors is possible if both species show overdispersed distributions throughout their host populations (Dobson, 1985).

Examples of competition from other sites within a fish are very scarce indeed but one such involving

* Tel: 01392 263757. Fax: 01392 263700.
E-mail: C.R.Kennedy@exeter.ac.uk

Parasitology (2001), **122**, S13–S22. Printed in the United Kingdom © 2001 Cambridge University Press

metacercarial stages of three species of diplostomatid digeneans in the eyes of perch, *Perca fluviatilis*, has been reported from a small lake, Slapton Ley, in Devon. The earliest study of the parasites of fish in this lake reported only *Diplostomum spathaceum*, although it was recognized that some individuals were found in the lens and others in the humour of the eye (Canning *et al.* 1973). A later study by Kennedy (1975) recognized the discrete identity of the individuals from the humour as belonging to the congeneric species *D. gasterostei*. This same paper also reported the appearance of metacercariae of *Tylodelphys clavata* in the humour of eyes of perch in October 1973, coincident with, and believed to be causally related to, the commencement of breeding by the Great Crested Grebe, *Podiceps cristatus*, in the lake. The population biology of *D. gasterostei* and *T. clavata* in perch in the lake was studied by Kennedy & Burrough (1977), who reported that the abundance of *T. clavata* increased rapidly whilst that of *D. gasterostei* remained relatively steady. A second species of *Tylodelphys*, *T. podicipina*, first appeared in the lake in 1976 and the metacercariae of this species also showed a preference for the humour of perch eyes (Kennedy, 1981*a*). As the abundance of this species increased, it became clear that that of *D. gasterostei* was simultaneously declining (Kennedy, 1981*b*). Two processes were believed to be involved in this decline: a decrease in the proportion of fish harbouring heavy infections of *D. gasterostei* which was significantly negatively correlated with the increase in population abundance of *T. clavata* in the eyes of perch and a dramatic decline in the rate of recruitment of *D. gasterostei* into young perch in 1978 and 1979, which coincided with the increase in infrapopulation density of *T. podicipina* in perch. It was thus suggested that the two species of *Tylodelphys* were interacting negatively with *D. gasterostei*, to the detriment of this latter species.

It was not possible to test this hypothesis directly or experimentally for a number of reasons. Nevertheless, interspecific competition was accepted as the most plausible explanation by Dobson (1985) who cited this as an example of exploitation competition. Meanwhile, the long-term monitoring of the helminth populations in perch, begun in 1973, was continued and a further paper on *T. podicipina* only appeared (Kennedy, 1987). In the winter of 1984/85 the lake froze over, winterkill caused a massive crash in the populations of perch, roach, *Rutilus rutilus*, and rudd, *Scardinius erythrophthalmus*, and their parasites, and hardly any fish were caught until 1990 (Kennedy, 1996; Kennedy, Wyatt & Starr, 1994). This provided an excellent opportunity to test the conclusions reached on the basis of the earlier pre-1985 studies, i.e. it represented a natural experiment. The recovery of the parasite populations was therefore monitored very closely up to the present (Kennedy, 1998).

The result is a 29 year data set, comprising samples taken over 15 years before the crash, intermittently during the crash and over a period of 10 years since the crash. The changes in population levels of the three species in the eyes of perch up to 1979 have already been published (Kennedy & Burrough, 1977; Kennedy, 1981*a*, *b*, 1987), but are also summarized here for clarity and to provide a comprehensive account. Changes in the population levels in the years immediately preceding the crash, and during and after the crash are the subject of this present paper, which also focuses on the decline in population size of *D. gasterostei* over the whole 29 year period. Additional data on the site of the three species within the humour of the eyes of perch are also presented. The overall aim is to test the hypothesis that interspecific competition occurs between the two species of *Tylodelphys* and *D. gasterostei* by examining the changes in the abundance of the three species from 1980 onwards and comparing these with the population changes before the crash with a view to determining if the same patterns are repeated over the period of recolonization.

MATERIALS AND METHODS

The study locality, Slapton Ley, Devon (NGR SX825440) is a small, shallow, eutrophic, freshwater coastal lake. Brief descriptions of the catchment and basins are given by Burt & Heathwaite (1996) and Johnes & Wilson (1996). A recent history of the fishery is given by Kennedy *et al.* (1994) and Kennedy (1996). The parasites all occurred as natural infections in the eyes of perch. The definitive host of *T. clavata* and *T. podicipina* is believed to be *Podiceps cristatus*: that of *D. gasterostei* is unknown. The first intermediate host of *T. clavata* and *T. podicipina* is *Physa fontinalis*, whereas that of *D. gasterostei* is *Lymnaea pereger*. Samples were taken at least once every year, but the frequency was variable depending upon personnel availability and funding. In some years samples were taken every month and in others only once or twice a year, but every effort was made to obtain a sample in October or the nearest month to this at the conclusion of the annual recruitment period. Wherever possible, a minimum of 30 fish was examined but variation in numbers and year class abundance between samples was extensive and it was never possible to obtain consistent or replicable samples. Fish were caught in gill nets or seine nets or by electric fishing as appropriate and were generally killed and deep frozen prior to examination. Fish required for a study of parasite site location were returned to the laboratory alive, and the eyes were frozen immediately after they were killed. All fish were weighed and measured and the majority were aged. Eyes were removed, opened from the site of entry of the optic nerve and searched, and all parasites removed, identified and counted. No significant difference in abundance between left and right eyes was ever found and so data were combined for each fish. Eyes required for a study of site selection were sectioned. All raw data sets will be deposited in the archives of the Slapton Ley Field Centre, operated by the Field Studies Council.

All terminology is used in accordance with the definitions of Bush *et al.* (1997). Data are presented in several ways, including abundance and prevalence levels of a species throughout the entire life span of a year class (cohort) of a fish (the year class is always the year of birth), and for fish of each year class at ages 0 + and 1 +. All such data are presented for samples taken in October or the nearest month to it to avoid problems of seasonality and ensure that the recruitment period for the year had concluded (Kennedy, 1981 *a*, *b*). This paper focuses on overall trends in parasite infrapopulation sizes and these are generally evident despite heterogeneity in sample composition and overdispersion. Other patterns are nested within the data set, but they are not dealt with here. Details of infrapopulation levels pre-1980 can be found in Kennedy (1981 *a*, *b*).

RESULTS

Seasonality

From 1973 to 1984, i.e. before the perch population crash, all three species showed a seasonal pattern of recruitment into perch. Both *T. clavata* and *D. gasterostei* infected fish of all sizes primarily in late summer with a second and minor infection period in spring. The life span of *T. clavata* is short, around 1 year only, and large numbers die over winter. Recruitment levels are higher each year of a fish's life and so the infrapopulation level increases over the life span of a fish whilst showing an autumn rise and winter decline. By contrast, *D. gasterostei* is long lived and although fish can be infected every year so that abundance rises over three or four years, it then declines in the oldest fish. Recruitment rates varied from year to year, but were consistently lower than those of *T. clavata* (Kennedy, 1981*b*). In complete contrast, *T. podicipina* can only infect young of the year perch in June and so fish are infected in one year of their life only. The parasites are long lived and infrapopulation densities decline throughout the life span of the host (Kennedy, 1987). These basic seasonal patterns in recruitment and mortality were also evident after the perch population crash.

The decline in D. gasterostei *infrapopulations*

The decline in infrapopulation levels of *D. gasterostei* was noted as early as 1975 (Kennedy & Burrough, 1977), when it was also reported that large numbers of *T. clavata* and *D. gasterostei* were seldom found in the same individual perch. The decline in recruitment of *D. gasterostei* was first noted in the 1978 and 1979 year classes (Kennedy, 1981*b*). Infrapopulation levels throughout the life-span of each year class of perch for the year classes 1976–1981 inclusive are shown in Table 1. For the year classes of 1976 and 1977, infection parameters of prevalence and abundance peaked in 0 + (prevalence) or 1 + (abundance) and declined thereafter, although the lower levels in older fish may also reflect the small sample sizes. Prevalence reached 100 % in 0 + fish of the 1976 cohort and 93 % in 0 + fish of the 1977 cohort, and abundance levels never fell below 3 (maximum 9·4) for fish up to 2 +.

However, maximum prevalence in 0 + fish of the 1978 year class was only 18 % and prevalence fell to 4·4 % in the 1981 year class (Table 1). Abundance levels in 0 + fish over the same period were also low and declined, from 0·28 to 0·04. Declines in prevalence throughout the life of each year class indicated that mortality exceeded recruitment, even though maximum abundance was often observed in older fish (3 + in the 1978 year class and 2 + in the 1980 and 1981 year classes). Over the period 1978–1981 abundance never exceeded 1 or variance 1·29, whereas between 1973 and 1977 abundance never fell below 3 (maximum 19·4) in fish up to 2 +, and variance frequently exceeded 100 (maximum 466). Clearly the declines in numbers of heavily infected fish and in recruitment rate noted for 1978 and 1979 continued through into the 1980 and 1981 year classes. Abundance never again reached the earlier levels, and in a year such as 1979 abundance in fish of that year class did not exceed 0·17 whereas abundance in fish of earlier year classes attained levels of up to 4·2 (3 + fish of the 1976 year class). Nevertheless, fish of all ages and year classes continued to harbour infections of *D. gasterostei*.

Population levels of all three species from 1980 until the crash

Levels of prevalence and abundance of *D. gasterostei* remained at this new low in the 1982, 1983 and 1984 year classes (Table 2). The two species of *Tylodelphys* maintained normal levels and seasonal patterns. Prevalence of *T. clavata* remained at or close to 100 % in both 0 + and 1 + fish, whilst abundance varied between good (1980, 1983) and poor (1981, 1985) recruitment years and was almost always higher in 1 + fish. The characteristic pattern of population biology and levels exhibited by *T. podicipina* in the 1970s was also unchanged. Prevalence approached or reached 100 % in 0 + fish of all year classes, whilst abundance remained at similar levels and was always highest in the 0 + fish. Prevalence declined only slightly in 1 + fish but abundance fell more sharply confirming that fish were infected on one occasion only.

Table 1. The decline in infection levels of *D. gasterostei* in perch from 1976 to 1984. Data are presented for the month of October only, or the month closest to that if no sample were taken in October

Perch year class			Perch age				
			0+	1+	2+	3+	4+
1976	Prevalence	%	100	96·2	94·1	81·8	58·3
	Mean abundance	$\bar{x}$	5·5	9·44	6·41	4·2	5·84
		S.D.	2·12	14·46	7·69	3·34	12·88
	Sample size	n	10	54	34	22	13
1977	Prevalence	%	93·3	92·9	73·1	18·9	33·3
	Mean abundance	$\bar{x}$	4·26	5·63	3·23	0·34	0·33
		S.D.	3·49	5·77	3·8	0·92	—
	Sample size	n	15	85	26	7	3
1978	Prevalence	%	18·0	16·7	8·0	3·6	5·0
	Mean abundance	$\bar{x}$	0·28	0·19	0·1	0·36	0·05
		S.D.	1·14	0·47	0·6	0·18	0·22
	Sample size	n	101	36	50	28	20
1979	Prevalence	%	16·9	2·8	0	16·6	—
	Mean abundance	$\bar{x}$	0·17	0·03	0	0·17	—
		S.D.	0·61	0·4	0	0·64	—
	Sample size	n	65	71	20	6	0
1980	Prevalence	%	0	4·2	16·7	7·6	—
	Mean abundance	$\bar{x}$	0	0·04	0·33	0·08	—
		S.D.	0	0·44	0·90	0·52	—
	Sample size	n	10	24	6	13	0
1981	Prevalence	%	4·4	0	10·0	0	—
	Mean abundance	$\bar{x}$	0·04	0	0·1	0	—
		S.D.	0·46	0	0·56	0	—
	Sample size	n	45	15	10	6	0

Thus, in the year classes from 1978 up to and including that of 1985 the population levels of *T. clavata* and *T. podicipina* remained more or less constant, whereas *D. gasterostei* remained at levels far lower than those reported pre-1978.

Population levels of all three species during the crash

The crash in the populations of perch, roach and rudd in the lake first became apparent in spring of 1985. Perch and parasite levels were normal in the last samples taken in 1984, but when sampling was resumed in 1985 no fish of any species were caught. Sampling was continued and intensified over the summer of 1985, but despite strenuous efforts only 12 perch, all 0+, were caught in the whole year (Table 2). The absence of *D. gasterostei* probably reflected the small sample size and low prevalence of the species at that time, as both *T. clavata* and *T. podicipina* were still present in the lake and infected 0+ fish of the 1985 year class. After 1985 and up to and including 1989 very few fish were caught and those very irregularly (Table 3). Samples were not in any way representative of year classes or fish ages, and any conclusions about parasite levels over this period must of necessity be very tentative. Nevertheless, the fact remains that the only species found throughout this period was *T. clavata*: neither *T. podicipina* nor *D. gasterostei* were recorded in any

sample. It is unlikely that their absence is simply a reflection of the small sample sizes and their irregular nature as 17 fish of the 1988 year class were caught in 1990 and 16 in 1991 but again only *T. clavata* was found in this larger sample of 33 fish. It would instead seem more likely that populations of *T. podicipina* and *D. gasterostei* had fallen below levels of detection or below the threshold levels at which transmission was possible and so had become locally extinct.

Population levels of all three species after the crash

Recovery began in 1990, with the capture of 2+ fish of the 1988 year class, and in 1991 more fish of this year class and some of the 1989 and 1990 year classes were also caught (Tables 2 and 3). There can be no doubt that *T. clavata* survived in the lake throughout the period of the crash and continued to infect fish of each year class. In 1991 it was found in 0+ perch, at levels of prevalence (100%) and abundance (39) comparable to those recorded before the crash (Table 2). Prevalence levels thereafter fluctuated between 100% and 61% in 0+ fish, and abundance between 30 and 2. In 1+ fish, prevalence was 100% in all years except 1996 and abundance varied from 280 to 27: the values of these parameters and their fluctuations were all within the ranges recorded for the pre-crash period (Table 2). By contrast, *T. podicipina*

Table 2. Infection levels of three species of digenean metacercariae in the eyes of perch from 1980–1999. Data are presented for the last month in each year in which a sample was taken and for the first two years of the fishes life only (–, no sample or parameter not calculated as sample too small)

	Perch year class														
	1980	81	82	83	84	85–*90	91	92	93	94	95	96	97	98	1999
D. gasterostei															
0+ %	0	4.4	4.7	9.1	2.0	0	9.0	0	0	0	5.5	0	–	0	16.1
0+ $\bar{x}$	0	0.04	0.05	0.14	0.02	0	0.09	0	0	0	0.05	0	–	0	0.22
0+ s.d.	0	0.5	0.45	0.67	0.01	0	1.7	0	0	0	0.5	0	–	0	0.8
0+ n	10	45	21	66	50	12	22	28	23	18	36	21	0	102	31
1+ %	4.2	0	20.8	–	0	–	0	0	0	0	0	0	0	0	
1+ $\bar{x}$	0.04	0	0.29	–	0	–	0	0	0	0	0	0	0	0	
1+ s.d.	0.4	0	0.78	–	0	–	0	0	0	0	0	0	0	0	
1+ n	24	15	24	0	3	0	37	13	9	17	4	13	2	14	
T. clavata															
0+ %	100	97.8	95.2	100	100	100	100	82.1	100	94.4	61.1	80.9	–	89.2	100
0+ $\bar{x}$	71.0	11.7	24.9	117.7	39.4	17.5	39	5.4	30.2	5.3	2.8	11.6	–	3.7	18.6
0+ s.d.	8.4	3.1	4.4	18.0	13.4	8.9	3.5	2.2	4.1	1.9	2.3	3.4	–	1.6	3.7
0+ n	0	45	21	66	50	12	22	28	23	18	36	21	0	102	31
1+ %	100	93.3	100	–	100	100	100	100	100	100	100	76.9	100	100	
1+ $\bar{x}$	71.8	11.3	329	–	77.0	39.0	119.7	171.4	116.2	280	165	27.6	41.5	87.1	
1+ s.d.	7.3	3.3	131	–	–	3.5	7.2	8.3	9.4	12.1	7.9	5.1	–	6.1	
1+ n	24	15	24	0	3	11	37	13	9	17	4	13	2	14	
T. podicipina															
0+ %	0	97.8	95.2	100	84.8	90.0	0	0	0	0	11.1	42.8	–	76.5	100
0+ $\bar{x}$	0	8.8	5.2	6.7	2.93	2.5	0	0	0	0	0.12	0.52	–	2.2	6.6
0+ s.d.	0	1.9	1.6	1.7	1.5	1.8	0	0	0	0	0.6	0.8	–	1.5	1.8
0+ n	10	45	21	66	50	12	22	28	23	18	36	21	0	102	31
1+ %	91.6	93.3	100	–	100	–	0	0	0	5.9	25.0	47.0	51.1	42.8	
1+ $\bar{x}$	3.37	5.6	3.25	–	2.0	–	0	0	0	0.06	0.25	0.76	1.02	0.7	
1+ s.d.	1.5	1.5	1.4	–	–	–	0	0	0	0.5	0.7	0.9	–	0.8	
1+ n	24	15	24	0	3	11	37	13	9	17	4	13	2	14	

* See Table 3.

Table 3. Infection levels of *T. clavata* in fish during the period of the population crash from 1986–1989. Data were obtained from fish caught in and after 1990 (–, no sample taken or parameter not calculated as sample too small). No *D. gasterostei* or *T. podicipina* were found

Age of fish			Year class of perch					
			1986	1987	1988*		1989*	
2+	Prevalence	%	–	–	100	100	100	100
	Mean abundance	$\bar{x}$	–	–	73	80·6	53·2	44·5
		S.D.	–	–	51·0	49·7	22·3	–
	Sample size	n	0	0	10	7	4	2
3+	Prevalence	%	–	100	100	100	100	
	Mean abundance	$\bar{x}$	–	29	69·9	79·3	152·5	–
		S.D.	–	–	45·6	53·7	46·9	–
	Sample size	n	0	2	10	6	4	0
4+	Prevalence	%	100		–	–	–	–
	Mean abundance	$\bar{x}$	91·5	–	–	–	–	–
		S.D.	–	–	–	–	–	–
	Sample size	n	2	0	0	0	0	0

* Fish of the 1988 and 1989 year classes were sampled in September/November 1990 and July/August 1991, hence the two columns for 1988 and 1989.

Table 4. Position of *T. clavata* and *T. podicipina* in the humour of perch eyes. (sus. lig = suspensory ligaments)

		Age of perch			
		1+	2+	3+	4+
T. clavata	Prevalence	100	100	100	100
	No. in humour (%)	17 (5·2)	254 (4·5)	262 (3·9)	16 (0·9)
	No. in sus. lig. (%)	308 (94·8)	5405 (95·5)	6376 (96·1)	16·71 (99·1)
	Total number	325	5659	6638	1687
	Intensity (S.D.)	32·5 (19·7)	202·1 (100·3)	442·5 (177·1)	562·3 (119·1)
T. podicipina	Prevalence	90	96·4	60	33·3
	No. in humour (%)	23 (46)	94 (62·7)	17 (94·4)	1 (100)
	No. in sus. lig. (%)	27 (54)	56 (37·3)	1 (5·6)	0
	Total number	50	150	18	1
	Intensity (S.D.)	5·6 (3·8)	5·6 (2·4)	2·0 (0·94)	1·0 (0)
	No. of fish examined	10	28	15	3

Data from Lamb (unpublished). NB. Only one perch (2 +) was found with *D. gasterostei*. This harboured 130 specimens, of which 128 were in the suspensory ligaments and 2 in the humour. It also harboured 19 *T. clavata* and 1 *T. podicipina* in the ligaments.

did not re-appear until 1995, and then in 1 + fish of the 1994 year class. Its population levels increased only slowly and pre-crash levels in 0+ perch of 100 % prevalence and 6·6 abundance were not attained until the 1999 year class: 6 years later. The recovery of *D. gasterostei* was very erratic: it was first recorded in 1991 but thereafter in some year classes only and only in 0+ perch (Table 2). Only in 1999, 8 years later, did it approach pre-crash (1978 and 1979) levels of 16·1 % prevalence and 0·22 abundance.

Site selection by the three species

Although all three species are generally described as living in the humour of the eye, as opposed to species such as *D. spathaceum* which inhabits the lens, they are in fact far more precise in their site selection than this. All three species show a preference for the suspensory ligaments (Table 4). Approximately 95 % of the metacercariae of *T. clavata* are found in the suspensory ligaments of young perch (Table 4) and this proportion increases to 99 % with fish age. By contrast, only 55 % of *T. podicipina* are found in the suspensory ligaments of young fish and this proportion declines to around 5 % in older fish as metacercariae disperse and move into the humour. Few data are available on *D. gasterostei* as this species was very uncommon at the time of this study in 1981, but such data as exist indicate that this species also prefers the suspensory ligaments. It can therefore be tentatively concluded that all three species have the same site preference.

DISCUSSION

The original hypothesis of the decline in *D. gasterostei* being due to negative interspecific interactions with both *T. clavata* and *T. podicipina* was based on: (1) Coincidences of timing such that (*a*) the decline in the proportion of perch with heavy infections of *D. gasterostei* coincided with increasing population levels of *T. clavata* after 1973, and (*b*) the dramatic decrease in the recruitment rate of *D. gasterostei* into young perch of the 1978 and 1979 year classes coincided with the appearance and population expansion of *T. podicipina* (Kennedy, 1981*a*, *b*, 1987). (2) A significant negative correlation between the decline in the proportion of perch carrying heavy infections of *D. gasterostei* (i.e. the % perch with 10 or more individuals) and the increase in maximum mean intensity of *T. clavata* (Kennedy, 1981*b*), such that heavy infections of *D. gasterostei* were seldom or never found in the same individual perch harbouring heavy infections of *T. clavata* (Kennedy & Burrough, 1977). There was no significant association (2×2 contingency test) between the presence and absence of the two species due to the very low number of fish harbouring neither parasite or *D. gasterostei* only.

The hypothesis was thus based on indirect evidence only, i.e. on field data on population changes. The hypothesis was supported by Dobson's (1985) independent analysis of the data as an example of exploitation competition, as he noted that the comparison between empirical data and his mathematical simulation was fairly striking.

The problem with many hypotheses of negative interspecific interactions is the difficulty of testing them. Hypotheses based on field data tend to be erected on the basis of observations on the decline in population levels of one species in the presence of another, as documented here, or on niche shifts (Chappell, 1969; Grey & Hayunga, 1980), or on exclusive distributions (Kennedy, Bates & Brown, 1989; Kennedy & Hartvigsen, 2000). All these observations are strongly suggestive of interspecific interactions, but confirmation needs to come from elsewhere and ideally from experiments. As Dobson (1985) has stated 'a variety of other biological and environmental factors contribute to the observed patterns of population behaviour'. Where it is possible to undertake experimental infections of fish with one or more species at different intensities, it can be shown that a change in abundance or niche shift by one species in the presence of another or even the exclusion of one species may be due to interspecific competition (Bates & Kennedy, 1990), but even then it may be difficult or impossible to determine whether the competition was for a nutrient resource or a more favourable site. In the present case of the metacercariae in the eyes of perch it was never possible to test the hypothesis of interspecific

competition experimentally in a laboratory. Because *T. podicipina* infects 0+ perch within a very short period after their birth (Kennedy, 1981*a*, 1987), it would have been necessary to obtain uninfected perch from the field and then maintain them in the laboratory for three or four months until cercariae of *D. gasterostei* and *T. clavata* were available for infection. This was not practicable, the more so as in the pre-crash period the identity of the first intermediate hosts of all three species was not known. The alternative approach of keeping the life cycle of all three species going in the laboratory was also impracticable in view of the problems of maintaining the bird hosts. Clearly, some other approach was required to provide confirmation of the hypothesis and/or to test it.

Routine monitoring of the perch and parasite populations up to 1984 confirmed that there had been no significant changes in the prevalences and abundances of any of the three species since 1979. The two species of *Tylodelphys* remained at their normal high levels, whilst *D. gasterostei* remained at its low post 1978 level and if anything declined even further. These observations served to extend the data set and confirm earlier observations e.g. that *D. gasterostei* rarely occurred in perch that harboured high numbers of either or both of the other two species. Further statistical analyses of the situation were not possible in view of the very low numbers of *D. gasterostei* and the absence of fish with no infections of either *Tylodelphys* species. The first intermediate host of each of the three species was identified, but unfortunately snail densities were never monitored. The numbers of *P. cristatus* increased and remained high until 1984/5 (Elphick, 1996) and both intermediate and definitive hosts appeared still to be available up to 1985. No obvious changes in any biological or environmental factors were identified that could provide an alternative or more satisfactory explanation than interspecific competition for the decline in *D. gasterostei*.

The investigation into the site location of the three species within the eye of perch provided some further support for the hypothesis of competition. The three species were found in the humour, in contrast to *D. spathaceum* which always occurred in the lens, and all three appeared to prefer the same site i.e. the suspensory ligaments. This is most evident in young fish. However, as fish aged and intensities of *T. clavata* increased, *T. podicipina* moved out of the ligaments and into the humour, whereas *T. clavata* remained in this former site. This could be interpreted as a niche shift, just as the spatial separation of the two species of *Diplostomum* could be considered an example of resource partitioning. Whatever the interpretation, there appears to be the potential for competition for the suspensory ligaments between the three species and for the humour between *D. gasterostei* and *T. podicipina*.

There appears to be no opportunity for resource partitioning in space or time for *D. gasterostei*. Wherever it attempts to locate in the eye, it will encounter one of the other three species. It can survive when alone in fish of any age, but in young fish it is likely to encounter high levels of *T. podicipina* and in older fish high levels of *T. clavata*. It has no refuge from the other two species, and such a situation is compatible with the competition hypothesis.

The matter might have rested there, with the hypothesis of interspecific competition still being preferred and indeed supported by new evidence but untested, were it not for the fish population crash over the winter of 1984/1985. The lake had become increasingly eutrophic throughout the 1970s and early 1980s, and in the very cold winter of 1984/1985 it froze over for a long period. This provided ideal conditions for winterkill (Kennedy *et al.* 1994; Kennedy, 1996), and perch, roach and rudd populations all crashed to virtually undetectable and unsampleable levels. It was never clear whether fish remained in the lake itself at densities too low for representative sampling, or whether they survived in a refuge close to the inflow of the main river where they could not be sampled. Over this period the numbers of *P. cristatus* declined (Elphick, 1996) presumably because of the decline in the numbers of fish. Recovery in the fish populations was slow: the recovery of rudd started in 1989 and that of roach and perch in 1990 (Kennedy, 1996). The grebes did not reach their pre-crash levels until 1990–1992 (Elphick, 1996). The virtual disappearance of roach and rudd and their parasites followed by their subsequent recovery were monitored very closely as they constituted a natural experiment for testing hypotheses erected from a study of their pre-crash populations (Kennedy *et al.* 1994; Kennedy, 1996). The crash in the perch population and the recovery and re-colonization by parasites provided a similar opportunity for testing the hypothesis of interspecific competition between the three species of eye flukes.

Both *D. gasterostei* and *T. podicipina*, together with other parasites of perch, (Kennedy, 1998) disappeared over the period of the fish crash. The most likely explanation is that transmission of each species became at best difficult and at worst impossible at such low densities of both perch and grebes. These two species were most vulnerable to any decline in perch numbers, since *D. gasterostei* densities were already very low before the crash and *T. podicipina* can only infect a single age group of perch, the 0 +.

Densities of all fish-eating birds, including grebes, declined during the crash period (Elphick, 1996) and it seems most probable that *T. podicipina*, and possibly also *D. gasterostei*, became temporarily locally extinct and re-colonized with the grebes. By 1991 the numbers of birds and perch were back to pre-crash levels. *Tylodelphys clavata* had remained in the lake and continued to transmit at pre-1985 rates throughout the period of the crash, and this almost certainly reflects the higher reproductive potential and recruitment rate of this species compared to the other two (Kennedy, 1981*b*). It appears that *D. spathaceum* also survived in the lake during the crash (Kennedy, 1998), and this suggests that the densities of the snail intermediate hosts did not decline significantly over the crash period.

In 1991, soon after the crash, *D. gasterostei* was present in the lake, snail, fish and bird hosts all appeared to have returned to pre-crash densities and a major competitor, *T. podicipina*, was absent and had not yet re-colonized. Why then, in the absence of *T. podicipina*, did *D. gasterostei* not increase in density to at least 1976 and 1977 levels as might have been predicted on the competition hypothesis? The most plausible explanation relates to the continued persistence of *T. clavata* at pre-crash levels over the whole period. This is the only species whose population densities have been shown statistically and significantly to be inversely correlated with those of *D. gasterostei*. It was evident (Kennedy & Burrough, 1977) that these two species could co-exist in the eyes of young fish, in which levels of *T. clavata* were lowest. It would therefore seem to be very significant that after 1990, *D. gasterostei* was only ever found in 0 + fish and typically in individual hosts in which levels of *T. clavata* were exceptionally low. Both species have the same preferred site in the eye and *T. clavata* appears able to retain occupation of it in the presence of either of the other species. If this interpretation is correct, it would suggest that *T. clavata* has a greater impact on *D. gasterostei* than does *T. podicipina*. This latter species has a much lower reproductive potential than its congener and is much slower to colonize, or re-colonize, a locality (Kennedy, 1981*a*, *b*; 1987).

Thus, the results of this natural experiment do confirm the results of the earlier study. These three species exhibit the same site preference within the eye and *T. clavata* appears able to retain its position there at the expense of the other two species. Why they prefer this particular site is unknown, but it is unlikely to be related to immune responses in any way. Dobson (1985) described the interaction as exploitation competition, but it is impossible at this point to identify what is being exploited and indeed the situation appears more akin to interference competition in many ways. The ability of *T. clavata* to emerge as the most successful competitor is entirely commensurate with its higher reproductive potential and so colonization ability (Kennedy, 1993, 1994). Persistence of all three species in perch is possible if all three species are overdispersed in distribution throughout the perch population (Dobson, 1985), and this has indeed been shown to be the case (Kennedy & Burrough, 1977; Kennedy,

1987). Thus the situation in respect of the eye flukes is comparable to that of other species of fish helminth identified as possible competitors. The extended data set and the recolonization of perch by the three species of eyefluke in Slapton Ley constituted a natural experiment that confirmed that negative interaction is the most parsimonius hypothesis that can be erected to explain the decline of *D. gasterostei* in the lake.

ACKNOWLEDGEMENTS

Over the period of 29 years too many people have helped in this study to be able to list them all individually. Nevertheless, I must single out P. C. and J. Shears and the Wardens and staff at the Slapton Ley Field Centre and thank them for their continual support. Financial support has been provided at various times by the NERC, SERC, Whitley Wildlife Conservation Trust and the Exeter University Research Fund. Thanks are also due to E. Lamb for permission to use unpublished data, and to everyone else who has assisted but has not been named.

REFERENCES

BATES, R. M. & KENNEDY, C. R. (1990). Interactions between the acanthocephalans *Pomphorhynchus laevis* and *Acanthocephalus anguillae* in rainbow trout: testing an exclusion hypothesis. *Parasitology* **100**, 435–444.

BATES, R. M. & KENNEDY, C. R. (1991). Potential interactions between *Acanthocephalus anguillae* and *Pomphorhynchus laevis* in their natural hosts chub, *Leuciscus cephalus* and the European eel, *Anguilla anguilla*. *Parasitology* **102**, 289–297.

BUCHMANN, K. (1988). Spatial distribution of *Pseudodactylogyrus anguillae* and *P. bini* (Monogenea) on the gills of the European eel, *Anguilla anguilla*. *Journal of Fish Biology* **32**, 801–802.

BURT, T. P. & HEATHWAITE, A. L. (1996). The hydrology of the Slapton catchments. *Field Studies* **8**, 543–557.

BUSH, A. O. & HOLMES, J. C. (1986). Intestinal helminths of lesser scaup ducks: an interactive community. *Canadian Journal of Zoology* **64**, 142–152.

BUSH, A. O., LAFFERTY, K. D., LOTZ, J. M. & SHOSTAK, A. W. (1997). Parasitology meets ecology on its own terms: Margolis *et al.* revisited. *Journal of Parasitology* **83**, 575–583.

CANNING, E. U., COX, F. E. G., CROLL, N. A. & LYONS, K. M. (1973). The natural history of Slapton Ley Nature Reserve: VI. Studies on the parasites. *Field Studies* **3**, 681–718.

CHAPPELL, L. H. (1969). Competitive exclusion between two intestinal parasites of the three-spined stickleback, *Gasterosteus aculeatus* L. *Journal of Parasitology* **55**, 775–778.

DOBSON, A. P. (1985). The population dynamics of competition between parasites. *Parasitology* **91**, 317–347.

ELPHICK, D. (1996). A review of 35 years of bird-ringing at Slapton Ley (1961–1995) together with a brief historical review of ornithological observations. *Field Studies* **8**, 699–725.

GREY, A. J. & HAYUNGA, E. G. (1980). Evidence for alternative site selection by *Glaridacris laruei* (Cestoida: Caryophyllidea) as a result of interspecific competition. *Journal of Parasitology* **66**, 371–372.

HALVORSEN, O. & MacDONALD, S. (1972). Studies on the helminth fauna of Norway XXVI: the distribution of *Cyathocephalus truncatus* (Pallas) in the intestine of brown trout (*Salmo trutta* L.). *Norwegian Journal of Zoology* **20**, 265–272.

HOLLAND, C. (1984). Interactions between *Moniliformis* (Acanthocephala) and *Nippostrongylus* (Nematoda) in the small intestine of laboratory rats. *Parasitology* **88**, 303–315.

HOLLAND, C. (1987). Interspecific effects between *Moniliformis* (Acanthocephala), *Hymenolepis* (Cestoda) and *Nippostrongylus* (Nematoda) in the laboratory rat. *Parasitology* **94**, 567–581.

HOLMES, J. C. (1973). Site selection by parasitic helminths: interspecific interactions, site segregation, and their importance to the development of helminth communities. *Canadian Journal of Zoology* **51**, 333–347.

JOHNES, P. J. & WILSON, H. M. (1996). The limnology of Slapton Ley. *Field Studies* **8**, 585–612.

KENNEDY, C. R. (1975). The natural history of Slapton Ley Nature Reserve VIII. The parasites of fish, with special reference to their use as a source of information about the aquatic community. *Field Studies* **4**, 177–189.

KENNEDY, C. R. (1981*a*). The establishment and population biology of the eyefluke *Tylodelphys podicipina* (Digenea: Diplostomatidae) in perch. *Parasitology* **82**, 245–255.

KENNEDY, C. R. (1981*b*). Long term studies on the population biology of two species of eyefluke, *Diplostomum gasterostei* and *Tylodelphys clavata* (Digenea: Diplostomatidae), concurrently infecting the eyes of perch, *Perca fluviatilis*. *Journal of Fish Biology* **19**, 221–236.

KENNEDY, C. R. (1987). Long term stability in the population levels of the eyefluke *Tylodelphys podicipina* (Digenea: Diplostomatidae) from perch. *Journal of Fish Biology* **31**, 571–581.

KENNEDY, C. R. (1992). Field evidence for interactions between the acanthocephalans *Acanthocephalus anguillae* and *A. lucii* in eels, *Anguilla anguilla*. *Ecological Parasitology* **1**, 122–134.

KENNEDY, C. R. (1993). Introductions, spread and colonization of new localities by fish helminth and crustacean parasites in the British Isles: a perspective and appraisal. *Journal of Fish Biology* **43**, 287–301.

KENNEDY, C. R. (1994). The ecology of introductions. In *Parasitic Diseases of Fish* (ed. Pike, A. W. & Lewis, J. W.), pp. 189–208. Tresaith, Dyfed, Samara Publishing Ltd.

KENNEDY, C. R. (1996). The fish of Slapton Ley. *Field Studies* **8**, 685–697.

KENNEDY, C. R. (1998). Aquatic birds as agents of parasite dispersal: a field test of the effectiveness of helminth colonisation strategies. *Bulletin of the Scandinavian Society for Parasitology* **8**, 23–28.

KENNEDY, C. R., BATES, R. M. & BROWN, A. F. (1989). Discontinuous distributions of the fish acanthocephalans *Pomphorhynchus laevis* and

Acanthocephalus anguillae in Britain and Ireland: an hypothesis. *Journal of Fish Biology* **34**, 607–619.

KENNEDY, C. R. & BURROUGH, R. (1977). The population biology of two species of eyefluke, *Diplostomum gasterostei* and *Tylodelphys clavata*, in perch. *Journal of Fish Biology* **11**, 619–633.

KENNEDY, C. R. & DI CAVE, D. (1998). *Gyrodactylus anguillae* (Monogenea): the story of an appearance and a disappearance. *Folia Parasitologica* **45**, 77–78.

KENNEDY, C. R. & GUÉGAN, J.-F. (1996). The number of niches in intestinal helminth communities of *Anguilla anguilla*: are there enough spaces for parasites? *Parasitology* **113**, 293–302.

KENNEDY, C. R & HARTVIGSEN, R. A. (2000). Richness and diversity of intestinal metazoan communities in brown trout *Salmo trutta* compared to those of eels *Anguilla anguilla* in their European heartlands. *Parasitology* **121**, 55–64.

KENNEDY, C. R., WYATT, R. J. & STARR, K. (1994). The decline and natural recovery of an unmanaged coarse fishery in relation to changes in land use and attendant eutrophication. In *Rehabilitation of Freshwater Fisheries* (ed. Cowx, I. G.), pp. 366–375. Oxford, Fishing News Books.

ROHDE, K. (1979). A critical evaluation of intrinsic and extrinsic factors responsible for niche restriction in parasites. *American Naturalist* **114**, 648–671.

ROHDE, K. (1991). Intra- and interspecific interactions in low density populations in resource-rich habitats. *Oikos* **60**, 91–104.

ROHDE, K. (1998). Is there a fixed number of niches for endoparasites of fish? *International Journal for Parasitology* **28**, 1861–1865.

STOCK, T. M. & HOLMES, J. C. (1987). *Dioecocestus asper* (Cestoda: Dioecocestidae): an interference competitor in an enteric helminth community. *Journal of Parasitology* **73**, 1116–1123.

STOCK, T. M. & HOLMES, J. C. (1988). Functional relationships and microhabitat distributions of enteric helminths of grebes (Podicipedidae): the evidence for interactive communities. *Journal of Parasitology* **74**, 214–227.

Concomitant infections, parasites and immune responses

F. E. G. COX*

Department of Infectious and Tropical Diseases, London School of Hygiene and Tropical Medicine, Keppel Street, London WC1E 7HT, UK

SUMMARY

Concomitant infections are common in nature and often involve parasites. A number of examples of the interactions between protozoa and viruses, protozoa and bacteria, protozoa and other protozoa, protozoa and helminths, helminths and viruses, helminths and bacteria, and helminths and other helminths are described. In mixed infections the burden of one or both the infectious agents may be increased, one or both may be suppressed or one may be increased and the other suppressed. It is now possible to explain many of these interactions in terms of the effects parasites have on the immune system, particularly parasite-induced immunodepression, and the effects of cytokines controlling polarization to the Th_1 or Th_2 arms of the immune response. In addition, parasites may be affected, directly or indirectly, by cytokines and other immune effector molecules and parasites may themselves produce factors that affect the cells of the immune system. Parasites are, therefore, affected when they themselves, or other organisms, interact with the immune response and, in particular, the cytokine network. The importance of such interactions is discussed in relation to clinical disease and the development and use of vaccines.

Key words: Concomitant infections, cytokines, protozoa, helminths, bacteria, viruses.

INTRODUCTION

The term concomitant infections, alternatively called mixed infections, traditionally refers to a situation in which two or more infectious agents coexist in the same host. In the light of modern concepts of biology this definition is insufficiently precise and the definition that will be used here is one in which the two (or more) concomitant infectious agents are specifically designated as being genetically different. This definition permits the inclusion of agents belonging to different species, the commonly accepted view of concomitant infections, and members of the same species that are genetically different, for example those belonging to a different strain or population. In nature, concomitant infections are the rule and this has been recognized since the earliest recorded times; for example, multiple infections of helminth eggs have been detected in human coprolites and other human remains from prehistoric sites (see Brothwell & Sandison, 1967; Cockburn, Cockburn & Reyman, 1998). What is less well known is that there are numerous interactions, both gross and subtle, between different kinds of organisms. This fact has been long recognized by experimental scientists, including parasitologists, who go to great lengths to use animals that are germ free, specific pathogen free (SPF) or harbour a known fauna or flora (gnotobiotic). Despite the widespread acceptance that different organisms commonly occurring together in the same hosts can, and do, influence one another directly or indirectly,

field workers and other parasitologists seldom consider more than the single organism that directly concerns them. The standard parasitological text books are silent on this subject and there is virtually nothing about concomitant infections in the more specialized texts on epidemiology (Anderson & May, 1991; Grenfell & Dobson, 1995; Isham & Medley, 1996), parasitism and host behaviour (Barnard & Behnke, 1990; Beckage, 1997), helminth infections and nutrition (Stephenson, 1987), evolution (Brooks & McLennan, 1993), immunology (Wakelin, 1996) or even host-parasite relationships (Toft, Aeschlimann & Bolis, 1991). There is some tangential reference to hormonal changes induced by parasites and possible effects on other parasites by Hillgarth & Wingfield (1995) but this is not pursued in any depth. There is, therefore, a major gulf between the well defined world of text book parasitology, with everything laid out in neat self-contained sections, and the nicely controlled conditions that exist in a laboratory, where there is usually a one to one parasite-host relationship, and the real world in the field where there may be many infections interacting with one another. There are many examples of concomitant infections in humans and animals (see for example Christensen *et al.* 1987; Ashford, 1991; Petney & Andrews, 1998; Viera *et al.* 1998). The infectious agents concerned may be those of the same species, related species or distantly related species. Among the best known examples of the interactions between parasites of the same species are the schistosomes where the presence of an ongoing infection of adult worms inhibits the establishment of a subsequent infection by larval forms, a phenomenon known as concomitant im-

* Tel: +44 020 7927 2333, Fax: +44 020 7580 9075.
E-mail: f.cox@lshtm.ac.uk

Parasitology (2001), **122**, S23–S38.   Printed in the United Kingdom   © 2001 Cambridge University Press

munity (Smithers, Terry & Hockley, 1969). This phenomenon is also seen in other helminth infections, for example, cestodes (see Heath, 1995). In malaria, an ongoing infection is thought by many workers not only to induce, but also to be necessary for immunity to a superimposed infection of parasites with the same or different genotype, a phenomenon called premunition (Sergent, 1937; Smith *et al.* 1999). Well known examples of interactions between more distantly related organisms are those that exist between the Epstein Barr virus and malaria parasites (Burkitt, 1969) and between the Human Immunodeficiency Virus (HIV) and several parasites of which the best known examples are *Cryptosporidium* and *Leishmania* spp. (see Ambroise-Thomas, this supplement).

One of the reasons why so little attention has been paid to concomitant infections is that the interactions involved are complex and difficult to understand. Briefly, such interactions can either be ecological, in which case the rules of ecology, particularly competition for space or resources, apply or immunological where the rules immunology apply. Anderson (1994) has stated, in another context, that 'the interaction between the variables that determine the typical course of infection in an individual patient and those that determine the typical course of infection in communities of people is often complex and very non-linear in form'. Everything that is said and implied here is made even more complex in the case of mixed infections. A number of ecologists, particularly those working with helminths, have begun to appreciate and take cognizance of mixed infections and their implications and a considerable amount of progress has been made in this area (Dobson, 1985, 1990; Poulin, 1998) and some of these aspects will be discussed further in this supplement by Dobson, Poulin, and Kennedy. However, the more subtle interactions that occur in hosts co-infected with more than one infectious agent, particularly those involving immunological responses, have been less well investigated and this is the area that will be considered in this article.

THE NATURE OF THE INFECTED HOST

It is a truism that a host harbouring any infectious agent is not the same as one that is not infected. Furthermore, hosts harbouring large numbers of parasites are not the same as those harbouring small numbers, those harbouring viruses are not the same as those harbouring bacteria and so on. It is not appropriate here to discuss in detail the nature of all the host's immune responses to infectious agents but it is important to appreciate the significance of T lymphocyte subsets and the cytokine network.

Essentially, from the moment an immunologically intact host is infected with any infectious agent, the host begins to mount an appropriate protective immune response. The key cells are the Th (T helper) lymphocytes. At first these cells are uncommitted but they gradually differentiate into Th_1 and Th_2 cells, each characterized by the cytokines they produce, until eventually they become fully differentiated and, when this happens, they are mutually exclusive. The Th_1 cells produce T_1 cytokines, particularly IL-2 that drives the immune response towards the production of cytotoxic T (T_c) cells and IFN-γ that drives the immune response towards the activation of macrophages. Th_2 cells, on the other hand, produce IL-4, IL-5, IL-10 and IL-13 that lead to the activation of B cells and the subsequent production of antibody, and to the proliferation and differentiation of eosinophils. In shorthand, the Th_1 responses represent the cell-mediated arm of the immune response and the Th_2 responses represent the humoral arm. For further information, there is an excellent account of this subject in Klein & Hořejší (1997). T_c cells are ideally suited for the destruction of virus-infected cells, IFN-γ-activated macrophages are involved in the killing of intracellular pathogens and the antibody produced by B cells is most effective against extracellular pathogens such as helminths. The role of Th_1 and Th_2 cells in a number of infectious diseases is well discussed in the various contributions in Romagnani (1996) and, with particular relevance to helminths, by Pritchard, Hewitt & Moqbel (1997). Parasites are no different from other pathogens in that they inevitably induce some kind of immune response except that T_c cells are less involved in parasitic infections than in viral infections (see Cox & Wakelin, 1998 and Wakelin, 1996 for general accounts of the immune responses to parasites). In general, it is widely accepted that protective Th_1 responses predominate in infections caused by protozoa whereas Th_2 responses are more important in immunity to helminth infections. In addition, the mutual exclusivity mentioned above frequently results in extreme polarization in which one arm of the immune response is protective and the other counter protective. However, these are generalisations and the details of each individual immune response can differ from time to time or from stage to stage of an infection (Allen & Maizels, 1997).

The polarization of T cells towards cell-mediated or antibody-mediated responses does not depend entirely on the infectious agent involved but can be modulated by pre-existing factors including cytokines. For example, the presence of IL-12 drives the immune response towards the T_1 pole whereas the presence of IL-4 drives it towards the T_2 pole (Ma *et al.* 1997). Initial or subsequent polarization involves the interaction of a number of regulatory cytokines some of which act as growth and differentiation factors. What is important here is that these cytokines, and also effector molecules, act non-

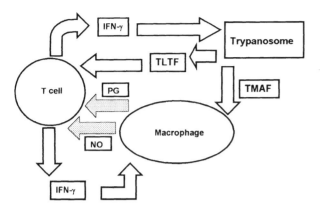

Fig. 1. Diagrammatic representation showing some of the interactions that exist between African trypanosomes and cytokine network. Trypanosomes produce a number of factors including trypanosome macrophage activating factor (TMAF) and trypanosome-derived lymphocyte triggering factor (TLTF) that stimulate macrophages and lymphocytes respectively. IFN-γ, a product of stimulated lymphocytes, can act as a trypanosome growth factor but prostaglandins (PG) and nitric oxide (NO) produced by macrophages inhibit lymphocyte activity and hence the production of IFN-γ. Trypanosomes are therefore enmeshed in a network of cytokines and effector molecules that both stimulate and inhibit their growth and development. In this diagram, positive signals are indicated by open arrows and inhibitory ones by stippled arrows.

specifically. It is therefore possible for any infectious agent to be caught up in the cytokine network. It is now becoming clear that this is what happens in the case of any agent acquired previously, concurrently or subsequently in the majority of concomitant infections at least under controlled laboratory conditions.

EFFECTS OF CYTOKINES AND PARASITE-DERIVED MOLECULES ON PARASITES

Although directed at other cells involved in the immune response certain molecules such as transforming growth factor-beta (TGF-β) and interferon-gamma (IFN-γ) can act directly or indirectly on parasites. For example, IFN-γ acts as a growth factor for *Trypanosoma brucei* (Bakhiet *et al.* 1996*b*) and TGF-β is required for the invasion of mammalian cells by *Trypanosoma cruzi* (Ming, Ewen & Pereira, 1995). Other examples are given by Barcinski & Costa-Moreira (1994) and Omer *et al.* (2000).

In addition to the molecules of the immune system, all parasites produce secreted or excreted products some of which can themselves affect cells of the immune system (see Kaye, 1999). The best studied molecules are those produced by African trypanosomes, particularly *T. brucei brucei*; trypanosome-derived lymphocyte triggering factor

(TLTF) and trypanosome macrophage activating factor (TMAF). TLTF induces lymphocytes to produce IFN-γ which both activates macrophages and promotes trypanosome growth and TMAF also stimulates macrophage activity (see Sternberg, 1998; Hamadien, Bakhiet & Harris, 2000). The production of TLTF is not restricted to *T. b. brucei* but has also been found in *T. b. gambiense*, *T. b. rhodesiense* and *T. evansi* (Bakhiet *et al.* 1996*a*). All the African trypanosomes are, therefore, potentially caught up in a series of interactions in which trypanosome-derived factors activate both T lymphocytes and macrophages. The net effect is that lymphocyte-produced IFN-γ enhances trypanosome growth while other trypanosome-derived molecules induce macrophages to produce molecules that inhibit lymphocyte activation (Fig. 1). Among other parasite-derived molecules there are some of protozoan origin that induce the synthesis of IL-12 by macrophages (Gazzinelli *et al.* 1997) and an IFN-γ-like molecule produced by the nematode *Trichuris muris* (Grencis & Entwistle, 1997). The precise role of these molecules is unclear but what is important here is that these parasite-derived factors can interact with the other molecules of the immune system and may be involved in enhancing or depressing the immune response to other organisms and cannot be ignored in any consideration of concomitant infections.

INTERACTIONS BETWEEN INFECTIOUS AGENTS

There have been numerous reports of interactions between parasites and between parasites and other infectious agents (see Christensen *et al.* 1987). This is not intended to be a comprehensive review of the subject and many of the examples cited by Christensen *et al.* will not be reiterated here. However, this is an attempt to collate and categorize some examples of the various interactions that have been recorded and to try to explain them in terms of what is happening in an infected host. In this respect, the Christensen *et al.* paper represents the end of an era. 1986 in an important watershed in our understanding of the interactions that occur between infectious agents because it was in that year that the Th$_1$/Th$_2$ dichotomy was suggested, first in mice (Mosmann & Coffman, 1989) then in humans (Romagnani, 1991) and subsequently in all species of mammals studied. This concept has dominated thinking about immune responses ever since (see Romagnani, 1996). Prior to this discovery, the various interactions between parasites and other infectious agents had been difficult to explain and mainly centred on attempts to implicate ill-defined mechanisms of immunodepression, often referred to as immunosuppression, brought about by molecules produced by parasites which facilitate their own survival. The best studied examples of immuno-depression are infections with African trypanosomes (Greenwood, 1974; Hudson & Terry, 1979), malaria parasites (Greenwood, 1974; Del Giudice, Grau & Lambert, 1988) and nematode worms (Behnke, 1987). According to the theory of immuno-depression, any concomitant infection that is acquired while the host is experiencing a phase of parasite-induced immunodepression should be able to establish itself more readily and possibly become more virulent or pathogenic. There is, for example, a correlation between the occurrence of Burkitt's lymphoma, caused by infection with the relatively harmless Epstein-Barr virus, and the presence of malaria (Burkitt, 1969). On the other hand, HIV does not seem to have much effect on malaria infections or vice versa (Butcher, 1992).

Not all factors that affect the outcome of con-comitant infections are immunological. Other factors responsible for a particular outcome include changes in the microenvironment. The induction of reticulocytosis in experimental rodent malaria infections, for example, should, and does, disad-vantage species such as *Plasmodium vinckei* and *P. chabaudi* that require mature erythrocytes resulting in lower parasitaemias. However, this explanation does not account for the fact that infections with *P. berghei*, which prefers reticulocytes, are similarly affected (see Cox, 1975a, 1978 and below).

These situations described above illustrate some of the problems inherent in seeking explanations for the interactions between the different agents in concomitant infections because there seem to be as many exceptions to the 'rules' as there are examples. It is, therefore, necessary to consider the nature of the possible outcomes. It is best to start with the simplest examples where a host that is harbouring, or has harboured, one organism A, is infected with a different organism, B. There are three outcomes with respect to B: the ensuing infection may be enhanced, suppressed or not affected in any way. However, there is also the reciprocal situation with respect to the infection caused by A, which may also be enhanced, depressed or neutral. Thus in this simple situation, there are a number of different possible outcomes. However, this example only takes into account a subsequent infection acquired at a specific point in time during the course of the ongoing infection. The outcome of infection B may well be different if the host harbouring A is infected with B at the beginning of the infection, at its peak or during a chronic phase. For example, the outcome of dual infections with of *Babesia microti* and *T. b. brucei* in mice varies according to when the piroplasm is administered in relation to the trypanosome (Millott & Cox, 1985). In this situation, the piro-plasm infection is always inhibited when the trypano-some is given beforehand but the inhibition increases with the interval between the two infections until at

−21 days it is absolute (Millott & Cox, 1985). In passing, it must be pointed out that these results are counterintuitive as it might be expected that super-imposing a piroplasm infection on an ongoing trypanosome infection, during a period of what should be trypanosome-induced immunodepression, should result in higher rather than lower piroplasm infections.

There are, therefore, many different outcomes resulting from concurrent infections and, from the examples cited above, it is clear that these may be unpredictable. In searching for the truth in any complex situation it is advisable to adopt the advice of William of Ockham (or Occam) as perpetuated in the maxim, Ockham's razor, which states that in seeking an explanation, various assumptions need not be multiplied needlessly; in other words, if there is a simple explanation there is no need to seek a more complicated one. However, this implies that all the knowledge necessary to produce a feasible explanation is available but, unfortunately, it is not. Therefore, the sensible approach is to seek simple explanations wherever possible, to look for exceptions and to modify the original explanation in terms of the exceptions and new knowledge.

In the sections below, selected examples of interactions between protozoa and viruses, protozoa and bacteria, protozoa and protozoa, protozoa and helminths, helminths and viruses, helminths and bacteria and between helminths and helminths will be discussed. It must be pointed out, however, that virtually all the information available comes from carefully controlled laboratory studies, usually with mice, that are sometimes very contrived. However, it is likely that these interactions will eventually be shown to apply to natural human and animal diseases and examples of such situations will also be discussed.

Protozoa and viruses

Viral infections are usually extremely difficult to detect and, therefore, the amount of information we have on interactions between protozoa and viruses is limited. Much of what we know comes from experiments involving *Plasmodium* spp. and other intraerythrocytic protozoa in mice and rats but there is no coherent pattern that emerges (see the reviews by Cox, 1975a, 1978). The intensity of *P. berghei* infections is suppressed in mice infected with West Nile virus (Yoeli, Becker & Bernkopf, 1955) or Newcastle Disease virus (Jahiel *et al.* 1968b) suggesting that virus-induced IFN-α might be protective against malaria parasites (Jahiel *et al.* 1968a). However, there is no direct evidence that this is the case but, in this context, it is interesting to note that, in humans, infections with *P. falciparum* are lower in patients co-infected with measles or influenza viruses (Rooth & Bjorkman, 1992). Thus viral infections may ameliorate malarial infections but, on the other hand, the non-lethal strains of *P. yoelii* and *P. chabaudi* become lethal in mice infected with the Rowson Parr virus (Cox, Wedderburn & Salaman, 1974) and infections with the Rowson Parr or urethane leukaemia virus also enhance *Babesia microti* infections in mice (Cox & Wedderburn, 1972). These results might be due to immuno-depression caused by the viruses which, as a group, tend to be immunosuppressive (Salaman, 1969) but these conflict with those cited above in which infections in virally infected animals or individuals were actually suppressed.

Turning now to other combinations of protozoa and viruses (other than HIV), there are a number of observations that can be attributed to the immuno-depressive effects of the virus infections and explained in terms of the consequent down regu-lation of cytokines required for immunity to the protozoan. *Cryptosporidium parvum* infections are enhanced in mice experimentally infected with the murine leukaemia retrovirus LP-BM5 (Darban *et al.* 1991) and this enhancement correlates with decreased IFN-γ and IL-2 production in the virus-infected mice (Alak *et al.* 1993). In naturally infected chickens, infections with *C. baileyi* are enhanced in animals co-infected with the chicken anaemia virus, CAV (Hornok *et al.* 1998). *Trypanosoma cruzi* infections are also more severe in mice co-infected with viruses; for example, infections with the murine leukaemia virus, MuLV, results in the enhancement of *T. cruzi* infections in mice (Silva *et al.* 1993) and this is also the case in mice co-infected with the mouse hepatitis virus type 3 (Verinaud *et al.* 1998). In the MuLV-infected mice, T cells do not respond to trypanosome antigens, suggesting immuno-depression on the part of the virus (Silva *et al.* 1993), and this is also the proposed explanation of the observations seen in the mouse hepatitis-infected animals (Verinaud *et al.* 1998).

There are very few studies on the effects of protozoan parasites on infections caused by viruses but those that have been described all indicate that viral infections are enhanced in animals harbouring parasitic protozoa. For example, infections with the murine oncogenic viruses, Murine Sarcoma virus or Moloney virus, are more severe in mice co-infected with *P. yoelii* (Salaman, Wedderburn & Bruce-Chwatt, 1969; Wedderburn, 1970, 1974) and mice co-infected with *T. cruzi* and MuLV develop a murine form of AIDS which does not occur in animals infected with either of these agents alone (Silva *et al.* 1993). The best documented evidence that protozoan infections can enhance viral infections in humans is that relating to the Epstein Barr virus which normally causes mild or inapparent infections but can contribute to the development of Burkitt's lymphoma in individuals exposed to malaria (see De The, 1985). Immunodepression is characteristic of

malaria infections in mice and humans (McGregor & Barr, 1962, see also Houba, 1988) and the en-hancement of viral infections in co-infected animals can be explained in these terms. What is not so easy to explain, however, is the finding that the feline immunodeficiency virus (FIV) causes changes to the activity of macrophages but this does not affect co-infection with *Toxoplasma gondii* (Lin & Bowman, 1992). Other interactions between *T. gondii* and immunosuppressive viruses are reviewed by Lacroix *et al.* (1996).

Protozoa and bacteria

Most of the information we have about interactions between protozoa and bacteria comes from studies on infections with intraerythrocytic protozoa. Early studies were mainly concerned with the spirochete, *Borrelia duttoni* which has little effect on infections with *P. berghei* (Colas-Belcour & Vervent, 1954) and *B. hispanica* which causes a slight enhancement of *P. berghei* infections in rats (Sergent & Poncet, 1957). As mentioned above, rickettsiae have a dramatic effect on rodent malaria infections and there is a considerable literature on the effects of *Eperythrozoon coccoides* and *Haemobartonella muris*, which both cause anaemia, and the consequent reticulocytosis that should favour the development of parasites like *P. berghei* that preferentially invade reticulocytes. In fact, the reverse is the case and *P. berghei* infections are suppressed in mice co-infected with *E. coccoides* (Peters, 1965). Infections with *P. chabaudi* and *P. vinckei*, that preferentially invade mature erythrocytes, are milder in mice harbouring *E. coccoides* which is easier to explain (Cox, 1966). *Haemobartonella muris* has received less attention but rats co-infected with this organism and *P. berghei* harbour higher infections of the malaria parasite than do uncontaminated controls, which is what one might expect in hosts with an increased proportion of the preferred host cells (Hsu & Geiman, 1952; Smalley, 1975). The interactions between rickettsial infections and rodent malaria parasites are important because of serious problems inherent in interpreting results obtained in laboratory mice infected with *E. coccoides* (see Cox, 1978).

A well explored but essentially experimental aspect of the possible interactions between bacteria and protozoa arises from studies in the 1970s that demonstrated that BGC, *Corynebacterium parvum* and other bacteria and bacterial products protect mice against malaria parasites and piroplasms (see Cox, 1975*a*). In fact, the number of bacteria and bacterial products that protect mice non-specifically against blood parasites is very large (Cox, 1981) and it is very likely that, in the field, bacterial infections play an important role in modulating infections with intraerythrocytic protozoa. The most likely expla-nation of the protection is the production of mediators, probably tumor necrosis factor (TNF) and nitric oxide (NO) by macrophages. This topic is discussed more fully elsewhere in this volume by Clark.

There are very few reports of interactions between bacteria and protozoa in humans but there are a number of reports of enhanced bacterial infections, particularly in children, suffering from severe malaria. Increased prevalence and parasite densities of *P. falciparum* appear to correlate with pertussis in children in contrast to the suppression of parasitaemias seen in those suffering from viral infections mentioned above (Rooth & Bjorkman, 1992). It is also well known that bacterial pneumonia is sometimes associated with malaria (Bygbjerg & Lanng, 1982; Mabey, Brown & Greenwood, 1987; Walsh *et al.* 2000) and there has been a report of enhanced tuberculosis in a patient with malaria (Hovette *et al.* 1999) but the reasons for this are not known. Given the importance of malaria it is surprising that the interactions between this disease and bacterial infections has not received much attention even in epidemiological studies (see Greenwood, 1997).

Protozoa and other protozoa

There is a massive literature concerned with the various interactions reported between protozoa and it is only possible here to draw on a few selected examples. These interactions can be divided into two main groups, interactions between parasites belonging to the same species and interactions between different species ranging from closely related to distantly related forms. The phenomenon of premunition in malaria infections has already been touched on and will not be discussed further here as it is more fully explored by Smith *et al.* (1999). Mixed infections with different species of malaria parasites are common (see Tanner & Baker, 1999) and all four species that infect humans have been found in a single individual (Purnomo *et al.* 1999). Much of what we know about the situation in humans comes from cross-sectional surveys and these indicate the involvement of both immunological and density-dependent factors in the regulation of parasitaemias (Bruce *et al.* 2000). Such studies are at the descriptive stage and it is not possible to speculate on the mechanisms involved without entering into the massive literature concerned with malaria immunology. The concept of specific immunity underlies much of our understanding of the epidemiology and control of malaria infections and this assumption appeared to suffer a major setback when it was demonstrated that there was a certain degree of heterologous immunity between different species of malaria parasites in rodents (reviewed by Cox, 1978). This was later extended to the discovery of protective immunity

between different genera of intraerythrocytic protozoa, *Plasmodium* and *Babesia* (Cox, H. W. & Milar, 1968; Cox, F. E. G., 1968). Attempts were initially made to explain these findings in conventional immunological terms, such as the presence of cross-reacting antigens, but it soon became clear that what happens is that the superimposed infection becomes involved in non-specific responses involving a number of mediators of inflammation, a topic that is discussed in more detail elsewhere in this supplement by Clark. Heterologous immunity is not restricted to the parasites of rodents and there is also convincing evidence that there is cross-immunity between the malaria parasites of non-human primates (Voller, Garnham & Turner, 1966) and some evidence that infections with *P. vivax* might reduce the severity of *P. falciparum* infections in humans (Maitland, Williams & Newbold, 1997).

In addition to the blood parasites, there are other examples of unexpected interactions between related protozoa belonging to different species. For example in mice infected with the coccidians *Eimeria falciformis* and *E. pragensis*, in that order, the cyst output of the latter is enhanced but in the reverse order there is no such effect (Shehu & Nowell, 1998).

There are also examples of dramatic interactions between distantly related species of protozoa, the most studied being those involving trypanosomes and other blood parasites. Some of the earliest investigations were concerned with co-infections with *T. lewisi* and *P. berghei* in rats in which the parasitaemias due to both infections, particularly the trypanosomes, were enhanced (Hughes & Tatum, 1956; Shmeul, Golensa & Spira, 1975). Experiments in mice infected with *T. musculi* provided similar results in mice co-infected with *P. berghei* (Büngener, 1975), *P. yoelii* (Cox, 1975b) or *Babesia microti* (Cox, 1977). *Trypanosoma cruzi* infections are also enhanced in mice infected with *P. berghei* (Krettli, 1977). Although difficult to explain at the time, these results can now be explained in terms of immunologically significant molecules, such as IFN-γ, acting as trypanosome growth factors as has been discussed above. However, this cannot be the whole story as *T. b. brucei* infections are not enhanced in mice co-infected with *B. microti* whereas the babesial infections are (Millott & Cox, 1985). In rats, infection with *T. lewisi* enhances *Toxoplasma gondii* infections (Guerrero, Chinchilla & Abrahams, 1997), although the mechanism is not at all clear and could be due to immunodepression or to the effects of trypanosome-derived molecules.

There are a number of reports of reciprocal enhancements of protozoan infections in mice, for example, *P. yoelii* and *Leishmania mexicana* (Coleman, Edman & Semprevivo, 1988) and *P. yoelii* and *L. amazonensis* (Coleman, Edman & Semprevivo, 1989). *P. berghei* infections are enhanced in rats co-infected with *Toxoplasma gondii*

(Rifaat *et al.* 1984) but there is no information about any effects on the toxoplasma infection. There are also examples of other one-sided interactions. The cyst output in mice infected with the intestinal flagellates *Spironucleus muris* and *Giardia muris* is reduced in animals also infected with *B. microti*, *P. berghei* or *P. yoelii* but the blood infections are not affected (Brett & Cox, 1982). These authors attribute the reduction in cyst output to physiological changes in the gut rather than to any immunological factors and this also seems to apply in the case of mice co-infected with *G. muris* and *Trichinella spiralis* (Roberts-Thomson *et al.* 1976 and see below). Interactions between protozoan infections in humans are not at all easy to assess but, with the increased use of more sensitive and specific diagnostic methods, it is likely that in the future there will be a number of such reports.

Protozoa and helminths

Intuitively, it would seem unlikely that there could be any interactions between single celled protozoan and multicellular helminths, particularly as most of them occupy different sites in the body and elicit very different kinds of immune responses. In fact, in virtually all situations where protozoa and helminths occur together that have been investigated experimentally, there is some degree of interaction, sometimes very dramatic (Christensen *et al.* 1987; Chieffi, 1992; Petney & Andrews, 1998; see Behnke *et al.* this supplement) Some of these interactions are now beginning to be subjected to the kind of analysis involving Th$_1$ and Th$_2$ cells referred to at the beginning of this review and this topic is discussed elsewhere in this supplement with relevance to nematode worms in rodents (see paper by Behnke *et al.*). The most studied interactions are those between trypanosomes and the intestinal nematode worm *Trichinella spiralis*. Mice infected with *T. spiralis* experience considerably increased *Trypanosoma musculi* infections when the nematode infection is initiated 5–10 days before the trypanosome and this enhancement is still obvious after 45 days (Bell, Adams & Ogden, 1984*a*). Similar enhancement of the trypanosome infection occurs in strains of mice differing in resistance to *T. spiralis* suggesting that this phenomenon is a general one that overrides the inherent susceptibility or resistance to the nematode (Chiejna & Wakelin, 1984). There are also reciprocal effects and the expulsion of *T. spiralis* is inhibited by infections with *T. musculi* also indicating a suppressive effect on the host's immune response, this time in the other direction (Bell, Adams & Ogden, 1984*b*). Thus it would seem that, in this model of doubly-infected animals, both the trypanosome and the nematode benefit although the details of the actual outcome vary from strain to strain of mouse. However the trypanosome also reduces the fecundity

of the worms which, therefore, suffer from the relationship in a different way (Bell, Adams & Ogden, 1984*b*). Rats infected with *T. b. brucei* also fail to expel the nematode worm *Nippostrongylus brasiliensis* (Urquhart *et al.* 1973), as is the case in another combination of trypanosome and nematode, *T. b. brucei* and *Trichinella spiralis*, in mice where the presence of the trypanosome also has an inhibitory effect on immunity to the worm (Onah & Wakelin, 1999). These authors measured IFN-γ and IL-4 levels, markers for Th$_1$ and Th$_2$ responses respectively and found that in the doubly-infected animals levels of IFN-γ were increased and levels of IL-4 were reduced and they conclude that in the doubly-infected animals the immune response is biased to the T$_1$ pole thus inhibiting immunity to *T. spiralis* which relies on the activation of T$_2$ responses.

The findings that trypanosome infections are enhanced in animals co-infected with nematodes relate to laboratory systems but there is some evidence that in domesticated animals the outcome is likely to be similar. In a natural situation, N'dama cattle in The Gambia are more susceptible to infection with *T. congolense* or *T. vivax* if they are infected with trichostongyle nematodes (Dwinger *et al.* 1994) and sheep infected with *Haemonchus contortus* and *T. congolense* are able to tolerate either infection but not both (Goossens *et al.* 1997). However, it is impossible to generalise from these findings and those discussed above and conclude that in dual infections with nematodes and trypanosomes the trypanosome infection will always be enhanced. For example in mice infected with *Heligmosomoides polygyrus* and *T. musculi* the trypanosome infection is not enhanced (Bell, Adams & Ogden, 1984*a*).

Schistosome infections interact with a variety of protozoa, at least in experimental models (see Chieffi, 1992). Infections with the rodent malaria parasites are affected in different ways by the presence of *Schistosoma mansoni*. For example, in the vole *Microtus guentheri*, *P. berghei* infections are enhanced if the two infections are given within 2 weeks of each other but depressed if the malaria infection is given seven weeks after the schistosome infection when the immune response to the worm is most active (Yoeli, 1956). There are also reciprocal interactions and *P. yoelii* actually inhibits the development of schistosome granulomas in experimentally infected mice (Abdel-Wahab *et al.* 1974). There have been some attempts to explain the complex and interdependent interactions between schistosomes and malaria parasites in terms of the Th$_1$/Th$_2$ dichotomy. In mice infected with *S. mansoni*, infections with the malaria parasite *P. chabaudi* are enhanced in animals infected with the worm 8 weeks previously while Th$_2$ responses to soluble egg antigens are reduced for 4 weeks after the malaria infection (Helmby, Kullberg & Troye-Blomberg, 1998), a finding that also applies in other situations that will be discussed below.

Schistosomes also interact with a number of other protozoa and infections with *T. cruzi* are much more severe in mice previously infected with *S. mansoni* (Kloetzel, Faleiros & Mendes, 1971) as are infections with *Toxoplasma gondii* (Kloetzel *et al.* 1977). *Schistosoma mansoni*-infected mice also experience more intense infections with the intestinal protozoa *Entamoeba muris*, *Trichomonas muris* and *Spironucleus muris* (Higgins-Opitz *et al.* 1990) but not *Leishmania major* (Yoshida *et al.* 1999) although *L. infantum* infections are more severe in hamsters infected with *S. mansoni* (Mangoud *et al.* 1998). Turning to the effect of protozoa and schistosome infections, *L. infantum* in hamsters delays *S. mansoni* granuloma formation (Morsy *et al.* 1998), a finding similar to that recorded for malaria parasites discussed above although when mice infected with *S. mansoni*, and undergoing an ongoing Th_2-inducing infection, are infected with *Toxoplasma gondii*, an infection that is responsive to Th_1 cytokines, there is a considerable reduction in schistosome granuloma size but there is no evidence of uncontrolled toxoplasma replication (Marshall *et al.* 1999). Taken together, these results suggest that infections with protozoa that stimulate a Th_1 response actually downregulate Th_2 responses with the result that the development of schistosome granulomas, a Th_2 phenomenon, is inhibited.

There is some indication that schistosomes interact with protozoa in the field and, from epidemiological studies in Egypt, it appears that the schistosomes interfere with the acquisition of immunity to *Entamoeba histolytica* and/or *E. dispar* (Mansour *et al.* 1997).

The expulsion of nematode infections is inhibited in mice infected with blood parasites, for example in mice infected with *B. microti* or *B. hylomysci* and *Trichuris muris* the expulsion of the worm is delayed, the suggested explanation being the immunodepression induced by the piroplasms (Phillips & Wakelin, 1976) and mice infected with *Plasmodium berghei* suffer from prolonged infections with *Nippostrongylus brasiliensis* and the self-cure mechanism is suppressed (Modrić & Mayberry, 1994). These authors attribute the results to decreased eosinophil production in the doubly-infected mice. Infections with intraerythrocytic protozoa, therefore, have an adverse effect on concomitant nematode infections. On the other hand, *B. microti* infections are not enhanced or prolonged in mice co-infected with *Heligmosomoides polygyrus* (Behnke, Sinski & Wakelin, 1999).

Among other records of interactions between protozoa and helminths is the observation that infections of *Entamoeba histolytica* are enhanced in mice infected with *Syphacla obvelata* (Vinayak & Chopra, 1978). There have been conflicting results from studies involving *T. spiralis* and *Eimeria* spp. In one study in mice, the outcome of infections with *T. spiralis* and *E. vermiformis* or *E. pragensis* differed according to the eimerian species; *E. vermiformis* delayed the expulsion of the worm whereas *E. pragensis* did not and the replication of *E. vermiformis* was enhanced in the presence of the worm whereas that of *E. pragensis* was reduced (Rose, Wakelin & Hesketh, 1994). These authors explain the results in terms of inflammation and immunological responses. In rats infected with *T. spiralis* and *E. nieschulzi* there were decreases in the number of adult worms and muscle parasitism and also in the numbers of oocysts produced by the protozoan (Stewart, Reddington & Hamilton, 1980). It is possible to interpret these last results in terms of changes to the architecture of the gut as the worms reach the gut when it has been severely damaged by the eimerian. A number of other examples of interactions between *T. spiralis* and *Eimeria* spp. are cited and discussed by Rose *et al.* (1994) who conclude that it is not possible to predict the outcome of the interactions between these parasites. Staying with intestinal infections and another combination involving *T. spiralis* and the intestinal flagellate, *Giardia muris*, the output of protozoan cysts is decreased in doubly-infected mice but the worm infection is not affected (Roberts-Thomson *et al.* 1976). The authors attribute this decrease in cyst output to changes in the gut rather than to any immunological factors, a conclusion also reached in the experiments using *G. muris* and the intraerythrocytic protozoa mentioned above (Brett & Cox, 1982).

Among other interesting interactions between helminths and protozoa are those between *Taenia crassiceps* and *T. cruzi* where, if the two are given to mice together, there is a slight delay in the onset of the trypanosome parasitaemia whereas if the trypanosome infection is initiated later during the cestode infection there are decreases in the levels of both IFN-γ and IL-4 and susceptibility to the trypanosome infection is markedly enhanced. In mice infected with *T. spiralis* and *L. infantum* 7 days later, when IFN-γ levels are already elevated, the subsequent leishmania infection is milder and the parasite load is reduced (Rousseau *et al.* 1997) and in mice infected with *T. spiralis* and *T. gondii*, also 7 days later, the protozoan infection is also milder (Afifi *et al.* 1999). In these experiments infection with *T. gondii* also protects against the nematode.

Helminths and viruses

Although concurrent infections of helminths and viruses are common the literature on the subject is rather limited and it is interesting to note that there are very few reports on the effects of immunosuppressive viruses on the outcome of helminth infections (see Markell, John & Krotoski, 1999). The best studied examples are those of the human lymphotropic virus type 1 (HTVL-1) that enhances infections with *Strongyloides stercoralis* (see Genta & Walzer, 1989). There is little evidence to suggest that HIV and schistosomes interact with one another despite the fact that they occur together in many parts of the world. However, there is some evidence that human infections with the hepatitis B virus contribute to liver damage caused by *S. mansoni* (Strauss & Lacet, 1986). There is also evidence that the viral infection is actually enhanced by the presence of the schistosomes and that there is an association between hepatitis B and the severity of schistosomiasis in Brazil (Strauss & Lacet, 1986) and Egypt (Madwar, El Tahawy & Strickland, 1989). Taken together, these findings indicate that infections with helminths can increase the severity of viral infections.

Helminths and bacteria

Some of the earliest reports of interactions between bacteria and helminths are concerned with a phenomenon called 'prolonged septicemic salmonellosis' in which patients infected with *S. mansoni* experienced prolonged bacteraemias due to several species of *Salmonella*. This finding was subsequently extended to other bacteria and the term 'prolonged septicaemic enterobacteriosis' was coined for this condition (see Chieffi, 1992). It has been suggested that this phenomenon is due either to immunodepression induced by the schistosomes or by the provision of some sort of protection to the bacteria (see Chieffi, 1992). There are, however, interactions between *Escherichia coli* toxins and schistosomes in mice in which the lethality caused by both agents is increased (see Chieffi, 1992). Much of the current interest in the effects of bacterial infections on helminths is indirect and is concerned with what happens in individuals vaccinated against *Mycobacterium tuberculosis*, the causative agent of tuberculosis, with BCG, particularly the purified protein derivative (PPD). BCG stimulates Th_1 immune responses and there is some concern that this might bias the response away from protection which, in the case of the helminths, tends to be Th_2 dependent. In fact, there is no evidence that this happens but there is evidence that the reverse is the case and in children sensitized *in utero* with *S. haematobium* or *Wuchereria bancrofti* the bias is away from the Th_1 responses required for protection against mycobacterial infections (Malhotra *et al.* 1999). Similarly, there is evidence that infection with *Onchocerca volvulus* may have an inhibitory effect on the immune responses to *M. tuberculosis* or *M. leprae*, the causative agent of leprosy (Stewart *et al.* 1999). These authors also comment on the significance of these findings with reference to reports of increased incidences of lepromatous leprosy in individuals with onchocerciasis. *Fasciola*

hepatica is another helminth that suppresses the protective Th_1 immune response against a bacterial infection, in this case *Bordetella pertussis* (Brady *et al.* 1999). It would appear from these examples that helminth infections that induce protective Th_2 immune responses can also downregulate Th_1 responses and that this might bring about the exacerbation of concomitant infections or a failure to respond to vaccination in infections that are controlled by Th_1 responses.

Helminths and other helminths

There are numerous examples of interactions between helminth worms (see Christensen *et al.* 1987) and those involving nematodes of rodents are discussed elsewhere in this supplement by Behnke and others. The best studied example of interactions between helminths is the phenomenon of concomitant immunity in which the adults of an ongoing infection prevent the establishment of another infection with larvae of the same species by eliciting an immune response which the adult worms can evade but the new larvae cannot. This seems to be a common phenomenon in helminth infections and is best exemplified by schistosomes (Smithers, Terry & Hockley, 1969) and by *Echinococcus granulosus* (see Heath, 1995). These interactions between worms belonging to the same species will not be discussed further here. Different species of helminth do, however, frequently occur in the same host and can interact with one another. The ecological aspects of such concurrent infections are discussed elsewhere in this supplement by Poulin and some of the immunological aspects of interactions between nematode worms are discussed by Behnke and his colleagues. It is only possible to discuss a few examples of different kinds of interactions here. In mice concurrently infected with the nematodes *Trichuris muris* and *Heligmosomoides polygyrus* the trichurid infection is rejected more slowly than in animals harbouring this parasite alone and the authors consider that this is due to a raising of the immune threshold necessary for the expulsion of the worms (Behnke, Ali & Jenkins, 1984). However, in most other systems investigated, the superimposed infection is suppressed by the presence of the original agent. In mice infected with *S. mansoni* and the cestode *Mesocestoides corti*, there is a reduction in the number of *M. corti* tetrathyridia and in the intensity of infection in the animals harbouring dual infections (Chernin *et al.* 1988). In mice infected with *S. mansoni* and the cestodes *Hymenolepis diminuta* or *Rodentolepis microstoma* there is an accelerated expulsion of the cestodes (Andreassen, Odaibo & Christensen, 1990). In mice co-infected with *S. mansoni* and *T. muris*, the Th_2 response to schistosome eggs appears to be involved in the elimination of the nematode infection (Curry *et al.* 1995) and in mice infected with *S. mansoni* and *Strongyloides venezuelensis* there is a decrease in the numbers of the nematode which the authors attribute to the effects of the immune response on the migrating larvae (Yoshida *et al.* 1999). There is also evidence from the field that infection with *S. mansoni* is in some way protective against infection with the geohelminths *Ascaris lumbricoides*, *Trichuris trichiura* and hookworms (Chamone *et al.* 1990). Thus, from these few examples it is clear that, in most combinations, infection with one worm can protect against others but that there are some circumstances in which infection with one helminth can enhance infection with another. Dual infections may also have synergistic effects on the pathology of some infections, for example children co-infected with hookworms and *Trichuris trichiura* have significantly less haemoglobin than children harbouring only one of these nematodes (Robertson *et al.* 1992).

DISCUSSION

This review, which of necessity is very selective, is intended to indicate the range of interactions between infectious agents that might affect the outcome of parasitic infections. It should be clear that mixed infections are the rule in natural situations and that protozoan infections are affected by other protozoa, helminths, bacteria and viruses while helminth infections are affected by protozoa, bacteria, viruses and other helminths. In virtually every combination that has been examined one or other of the concomitant agents is affected by the presence of the other and in many cases both are affected. In addition to these infectious agents, there are also fungi and prions to consider and virtually nothing is known about how these interact with parasites. The outcome of any interaction is not necessarily predictable and can vary from stage to stage of the infection. The age and sex of the host, factors that have not been considered here, can also influence the outcome of an infection. The important thing to be borne in mind is that these interactions do occur and cannot be ignored.

Immunodepression

The most common outcomes in dual infections are that the infection caused by one or other of the agents may be enhanced or depressed, both may be affected, one may be enhanced and the other depressed or vice versa. In addition, the presence of one agent can increase or decrease the severity of the pathology caused by the other. With all these possibilities the outcomes of any combination might be unpredictable but there are a number of patterns that seem to be consistent and a beginning has been made in unravelling these patterns. The most common situation is one in which one agent causes

immunodepression and a superimposed infection is able to take advantage of this situation. Essentially this is no different from the situation in which many immunocompromised hosts are susceptible to intercurrent infections with a range of microorganisms. However, it is too simplistic to assume that any parasitic infection in an immunocompromised host is likely to be enhanced and, in this context, it is interesting to note that very few parasitic infections are actually significantly enhanced in individuals infected with HIV (see Ambroise-Thomas, in this volume). Some degree of immunodepression is common, if not universal, during the course of infections with parasites and microorganisms but the methods used to assess the immune status of the host need not necessarily reveal those aspects that are relevant to the superimposed infection. For example, many of the early observations on immunodepression in parasitic infections were based on counting antibody-producing cells, which is no longer appropriate given our present knowledge of the roles of T cells in immune responses. Immunodepression during parasitic infections may be due to by-products of an ongoing protective immune response or to factors induced by the parasites themselves in order to ensure their own survival or to damp down immunopathological changes. Various aspects of these processes are discussed in more detail in the various contributions in Doenhoff & Chappell (1997). What is important to understand here is what the essential elements of a particular immune response are that render the host more or less susceptible to infection. There may be other alternatives to explain the enhanced infections seen during concomitant infections such as factors that enhance the growth or development of the parasite produced as part of the normal immune response or alterations to the cells required for the survival of a particular parasite.

The Th_1/Th_2 dichotomy

The second important factor in determining the outcome of an infection is whether or not the established infection is inducing a Th_1 or a Th_2 response. The Th_1 immunological milieu involves a number of molecules and the cells that produce them, in particular, NK cells, IL-12 and IFN-γ. The initiation of a new immune response in such a situation is gradually forced towards the Th_1 pole and, if the superimposed agent is controlled by Th_2 responses, it is at an advantage in such a situation. The reverse applies if IL-4 predominates in the immunological milieu at the time of the second infection. There is now a considerable amount of solid work going on in this area and from the experiments that have been described it looks as if the pattern described above is of general application. However, it is important to bear in mind that the

Th_1/Th_2 paradigm is not a rigid set of rules (Allen & Maizels, 1997) and that there may be switches in the patterns of Th_1 and Th_2 cytokines during the course of an infection as has been clearly demonstrated in the case of rodent malarias (Taylor-Robinson & Phillips, 1998).

Immune effector molecules

A number of molecules involved in the immune response act on a variety of cells and can affect parasites directly. Mention has already been made of IFN-α which, although important in viral infections, does not seem to play any significant part in the immune responses induced by, or otherwise involved in, parasitic infections. There are, however, a number of molecules that are relevant, the most studied being the lymphocyte product, IFN-γ, and the macrophage products TNF, TGF-β, NO and reactive oxygen intermediates. Regardless of how a particular immune response was initiated these molecules act non-specifically and thus a super-imposed parasite may be affected by effector molecules the production of which it itself has not initiated, in other words, the bystander effect.

Parasite-derived immunomodulatory factors

It is now clear that a number of parasites produce molecules that affect cells of the immune response directly or indirectly and thus exert immuno-modulatory effects. The best characterized are trypanosome macrophage activating factor (TMAF) and trypanosome lymphocyte triggering factor (TLTF) and it is almost certain that a number of others will be described. The presence and effects of such molecules will have to be taken into account when considering concomitant infections.

The significance of studies on concomitant infections

It has been the purpose of this review to give some indication of the range of interactions that occur when a host is infected with more than one infectious agent. These interactions cover the whole spectrum from the enhancement to suppression of one or other or both of the co-infecting agents and from the augmentation to the amelioration of the pathology of the infection. This means that the nature of any specific parasitic infection in a host concurrently infected with another infectious agent may be very different from an infection caused by the same parasite in a host co-infected with another agent or in a host that is otherwise uninfected. This has a number of implications that relate to the epidemio-logical and clinical aspects of human and veterinary parasitology and to the development of vaccines and the use of chemotherapy. The best known example of the clinical implications of interacting infections is Burkitt's lymphoma, resulting from an interaction

between the Epstein-Barr virus and malaria (Burkitt, 1969; De The, 1985) where the host appears to lose the T cell control of the virus infection (Whittle *et al.* 1984). There are also other examples including the increased incidence of lepromatous leprosy in patients with onchocerciasis (Stewart *et al.* 1999) and more severe strongyloidiasis in patients harbouring the virus HTVL-1 (Genta & Walzer, 1989). There is now a vast amount of evidence from experimental studies to suggest that the clinical picture in many infections may be markedly influenced by the presence of unrelated organisms and it is probable that such influences will prove to be the rule rather than the exception.

Concomitant infections are also likely to affect the efficacy of vaccines and will influence the design of new vaccines. It is already known that it is difficult to vaccinate children with malaria against tetanus, typhoid or bacterial meningococcus (Williamson & Greenwood, 1978, and see Björkman, 1988). Children sensitized to helminth antigens also appear to have an impaired response to mycobacterial antigens (Malhotra *et al.* 1999). New antiparasitic vaccines will have to take account of the possibility of inhibiting the immune responses to antimicrobial and antiviral vaccines by triggering inappropriate immune responses. Such inappropriate responses might also result from chemotherapy which causes the death of the parasites and the release of sequestered antigens as occurs in some autoimmune diseases. It has been suggested that new anti-parasitic vaccines should be designed to trigger the appropriate T_1 or T_2 response (Cox, 1997) but care will also have to be taken to ensure that the efficacy of such vaccines is not impaired by concomitant infections.

Concomitant infections have long been ignored by parasitologists but the time has now come to accept that such infections may be the rule and not the exception and to test laboratory findings in the field. Our understanding of the totality of the immune response, particularly its cytokine control, has now provided us with tools to analyse what is happening during the course of mixed infections and it would be remiss of us not to take advantage of these tools and to apply the findings to the rational control of parasitic infections in animals and humans.

REFERENCES

ABDEL-WAHAB, M. F., POWERS, K. G., MAHMOUD, S. S. & GOOD, W. C. (1974). Suppression of schistosome granuloma formation by malaria in mice. *American Journal of Tropical Medicine and Hygiene* **23**, 915–918.

AFIFI, M. A., EL-HOSEINY, L. M., MOUSTAFA, M. A., GAMRA, M. M. & KHALIFA, K. E. (1999). Reciprocal heterologous protection between *Trichinella spiralis* and *Toxoplasma gondii* concurrently present in experimental murine models. *Journal of the Egyptian Society of Parasitology* **29**, 963–978.

ALAK, J. I., SHAHBAZIAN, M., HUANG, D. S., WANG, Y., DARBAN, H., JENKINS, E. M. & WATSON, R. R. (1993). Alcohol and murine immunodeficiency syndrome suppression of *Cryptosporidium parvum* infection during modulation of cytokine production. *Alcohol Clinical and Experimental Research* **17**, 539–544.

ALLEN, J. E. & MAIZELS, R. M. (1997). Th1–Th2: reliable paradigm or dangerous dogma? *Immunology Today* **18**, 387–392.

ANDERSON, R. M. (1994). Populations, infectious disease and immunity: a very nonlinear world. *Philosophical Transactions of the Royal Society, Section B* **346**, 457–505.

ANDERSON, R. M. & MAY, R. M. (1991). *Infectious Diseases of Humans. Dynamics and Control*. Oxford: Oxford University Press.

ANDREASSEN, J., ODAIBO, A. B. & CHRISTENSEN, N. Ø. (1990). Antagonistic effects of *Schistosoma mansoni* on superimposed *Hymenolepis diminuta* and *H. microstoma* infections in mice. *Journal of Helminthology* **64**, 337–339.

ASHFORD, R. W. (1991). The human parasite fauna: towards an analysis and interpretation. *Annals of Tropical Medicine and Parasitology* **85**, 189–198.

BAKHIET, M., BÜSCHER, P., HARRIS, R. A., KRISTENSSON, K., WIGZELL, H. & OLSSON, T. (1996*a*). Different *Trypanozoon* species possess CD8 dependent lymphocyte triggering factor-like activity. *Immunology Letters* **50**, 71–80.

BAKHIET, M., OLSSON, T., MHLANGA, P., BÜSCHER, P., LYCKE, N., VAN DER MEIDE, P. H. & KRISTENSSON, K. (1996*b*). Human and rodent interferon-gamma as a growth factor for *Trypanosoma brucei*. *European Journal of Immunology* **26**, 1359–1364.

BARCINSKI, M. A. & COSTA-MOREIRA, M. E. (1994). Cellular responses of protozoan parasites to host-derived cytokines. *Parasitology Today* **10**, 352–355.

BARNARD, C. J. & BEHNKE, J. M. (eds) (1990). *Parasitism and Host Behaviour*. London: Taylor and Francis.

BECKAGE, N. E. (ed.) (1997). *Parasites and Pathogens. Effects on Host Hormones and Behavior*. New York: International Thomson Publishing.

BEHNKE, J. M. (1987). Evasion of immunity by nematode parasites causing chronic infections. *Advances in Parasitology* **26**, 1–71.

BEHNKE, J. M., ALI, N. M. & JENKINS, S. N. (1984). Survival to patency of low level infections with *Trichuris muris* in mice concurrently infected with *Nematospiroides dubius*. *Annals of Tropical Medicine and Parasitology* **78**, 509–517.

BEHNKE, J. M., SINSKI, E. & WAKELIN, D. (1999). Primary infections with *Babesia microti* are not prolonged by concurrent *Heligmosomoides polygyrus*. *Parasitology International* **48**, 183–187.

BELL, R. G., ADAMS, L. S. & OGDEN, R. W. (1984*a*). *Trypanosoma musculi* with *Trichinella spiralis* or *Heligmosomoides polygyrus*: concomitant infections in the mouse. *Experimental Parasitology* **58**, 8–18.

BELL, R. G., ADAMS, L. S. & OGDEN, R. W. (1984*b*). *Trypanosoma musculi* and *Trichinella spiralis*: concomitant infections and selection for resistance genotypes in mice. *Experimental Parasitology* **58**, 19–26.

BJÖRKMAN, A. (1988). Interactions between chemotherapy and immunity to malaria. In *Malaria Immunology* (ed. Perlmann, P. & Wigzell, H). *Progress in Allergy* **41**, 331–356.

BRADY, M. T., O'NEILL, S. M., DALTON, J. P. & MILLS, K. H. (1999). *Fasciola hepatica* suppresses a protective Th$_1$ response against *Bordetella pertussis*. *Infection and Immunity* **67**, 5372–5378.

BRETT, S. J. & COX, F. E. G. (1982). Interactions between the intestinal flagellates *Giardia muris* and *Spironucleus muris* and the blood parasites *Babesia microti*, *Plasmodium yoelii* and *Plasmodium berghei* in mice. *Parasitology* **85**, 101–110.

BROOKS, D. R. & McLENNAN, D. A. (1993). *Parascript. Parasites and the Language of Evolution*. Washington: Smithsonian Institution Press.

BROTHWELL, D. & SANDISON, A. T. (eds) (1967). *Diseases in Antiquity*. Springfield, Illinois: C. T. Thomas.

BRUCE, M. C., DONNELLY, C. A., ALPERS, M. P., GALINSKI, M. R., BARNWELL, J. W., WALLIKER, D. & DAY, K. P. (2000). Cross-species interactions between malaria parasites in humans. *Science* **287**, 845–848.

BÜNGENER, W. (1975). Verlauf von *Trypanosoma musculi*-Infektion in mit *Plasmodium berghei* infizierten Mäusen. *Zeitschrift für Tropemedizin und Parasitologie* **26**, 285–290.

BURKITT, D. P. (1969). Etiology of Burkitt's lymphoma – an alternative hypothesis to a vectored virus. *Journal of the National Cancer Institute* **42**, 19–28.

BUTCHER, G. A. (1992). HIV and malaria: a lesson in immunology. *Parasitology Today* **8**, 307–311.

BYGBJERG, I. C. & LANNG, C. (1982). Septicaemia as complication of falciparum malaria. *Transactions of the Royal Society of Tropical Medicine and Hygiene* **76**, 705.

CHAMONE, M., MARQUES, C. A., ATUNCAR, G. S. & PEREIRA, A. L. A. (1990). Are there interactions between schistosomes and intestinal nematodes? *Transactions of the Royal Society of Tropical Medicine and Hygiene* **84**, 557–558.

CHERNIN, J., McLAREN, D. J., MORINAN, A. & JAMIESON, B. N. (1988). *Mesocestoides corti*: parameters of infection in CBA/Ca mice and the effects of introducing a concomitant trematode infection. *Parasitology* **97**, 393–402.

CHIEFFI, P. P. (1992). Interrelationship between schistosomiasis and concomitant disease. *Memórias do Instituto Oswaldo Cruz* **87**, 291–296.

CHIEJINA, S. N. & WAKELIN, D. (1984). Interactions between infections with blood protozoa and gastrointestinal nematodes. *Helminthologia* **31**, 17–21.

CHRISTENSEN, N. Ø., NANSEN, P., FAGBEMI, B. O. & MONRAD, J. (1987). Heterologous antagonistic and synergistic interactions between helminths and between helminths and protozoans in concurrent experimental infection of mammalian hosts. *Parasitology Research* **73**, 387–410.

COCKBURN, A., COCKBURN, E. & REYMAN, T. A. (eds) (1998). *Mummies, Disease and Ancient Cultures*. 2nd edn. Cambridge: Cambridge University Press.

COLAS-BELCOUR, J. & VERVENT, G. (1954). Sur des infections mixtes de la souris à spirochétes récurrents et *Plasmodium berghei*. *Bulletin de la Société de Pathologie Exotique* **47**, 493–497.

COLEMAN, R. E., EDMAN, J. D. & SEMPREVIVO, L. H. (1988). *Leishmania mexicana*: effect of concomitant malaria on cutaneous leishmaniasis. Development of lesions in a *Leishmania*-susceptible (BALB/c) strain of mouse. *Experimental Parasitology* **65**, 269–276.

COLEMAN, R. E., EDMAN, J. D. & SEMPREVIVO, L. H. (1989). The effect of pentostam and cimetidine on the development of leishmaniasis (*Leishmania mexicana amazonensis*) and concomitant malaria (*Plasmodium yoelii*). *Annals of Tropical Medicine and Parasitology* **83**, 339–344.

COX, F. E. G. (1966). Acquired immunity to *Plasmodium vinckei* in mice. *Parasitology* **56**, 719–732.

COX, F. E. G. (1968). Immunity to malaria after recovery from piroplasmosis in mice. *Nature* **219**, 646.

COX, F. E. G. (1975 a). Factors affecting infections of mammals with intraerythrocytic protozoa. *Symposia of the British Society for Experimental Biology* **29**, 429–451.

COX, F. E. G. (1975 b). Enhanced *Trypanosoma musculi* infections in mice with concomitant malaria. *Nature* **258**, 148–149.

COX, F. E. G. (1977). Interactions between trypanosomes and piroplasms in mice. *Protozoology* **3**, 129–134.

COX, F. E. G. (1978). Concomitant infections. In *Rodent Malaria* (ed. Killick-Kendrick, R. & Peters, W.), pp. 309–342. London: Academic Press.

COX, F. E. G. (1981). Non-specific immunization against parasites. In *Isotopes and Radiation in Parasitology IV*, pp. 91–100. Vienna: International Atomic Energy Agency.

COX, F. E. G. (1997). Designer vaccines for parasitic diseases. *International Journal for Parasitology* **27**, 1147–1157.

COX, F. E. G. & WAKELIN, D. (1998). Immunology and immunopathology of human parasitic infections. In *Topley and Wilson's Microbiology and Microbial Infections*, 9th edn. Volume 5, *Parasitology* (ed. Cox, F. E. G., Kreier, J. P. & Wakelin, D.), pp. 57–84. London: Arnold.

COX, F. E. G. & WEDDERBURN, N. (1972). Enhancement and prolongation of *Babesia microti* infections in mice infected with oncogenic viruses. *Journal of General Microbiology* **72**, 79–85.

COX, F. E. G., WEDDERBURN, N. & SALAMAN, M. H. (1974). The effect of Rowson-Parr Virus on the severity of malaria in mice. *Journal of General Microbiology* **85**, 358–364.

COX, H. W. & MILAR, R. (1968). Cross-protection immunization of *Plasmodium* and *Babesia* infections of rats and mice. *American Journal of Tropical Medicine and Hygiene* **17**, 173–179.

CURRY, A. J., ELSE, K. J., JONES, F., BANCROFT, A., GRENCIS, R. K. & DUNNE, D. W. (1995). Evidence that cytokine-mediated immune interactions induced by *Schistosoma mansoni* alter disease outcome in mice concurrently infected with *Trichuris muris*. *Journal of Experimental Medicine* **181**, 769–774.

DARBAN, H., ENRIQUEZ, J., STERLING, C. R., LOPEZ, M. C., CHEN, G., ABBASZADEGAN, M. & WATSON, R. R. (1991). Cryptosporidiosis facilitated by retroviral infection with LP-BM5. *Journal of Infectious Diseases* **164**, 741–745.

DEL GIUDICE, G., GRAU, G. E. & LAMBERT, R. H. (1988). Host responsiveness to malaria epitopes and immunopathology. In *Malaria Immunology* (ed. Perlmann, P. & Wigzell, H). *Progress in Allergy* **41**, 288–330.

De THE, G. (1985). Epstein-Barr virus and Burkitt's lymphoma worldwide; the causal relationship revisited. In *Burkitt's Lymphoma: A Human Cancer Model* (ed. Lenoir, G. M., O'Conor, G. T. & Olweny, C. L. M.), pp. 165–176. Lyon: IARC Scientific Publications No. 60.

DOBSON, A. P. (1985). The population dynamics of competition between parasites. *Parasitology* **91**, 317–347.

DOBSON, A. P. (1990). Models for multi-species parasite-host communities. In *Parasite Communities. Patterns and Processes* (ed. Esch, G., Bush, A. & Aho, J.), pp. 261–288. London: Chapman and Hall.

DOENHOFF, M. J. & CHAPPELL, L. H. (eds) (1997). Survival of parasites, microbes and tumours: strategies for evasion, manipulation and exploitation of the immune response. *Parasitology* **115** (Suppl.). S1–S175.

DWINGER, R. H., AGYEMANG, K., KAUFMANN, J., GRIEVE, A. S. & BAH, M. L. (1994). Effects of trypanosome and helminth infections on health and production parameters of village N'Dama cattle in the Gambia. *Veterinary Parasitology* **54**, 353–365.

GAZZINELLI, R. T., CAMARGO, M. M., ALMEIDA, I. C., MORITA, Y. S., GIRALDO, M., ACOSTA-SERRANO, A., HEINY, S., ENGLUND, P. T., FERGUSON, M. A. J., TRAVASSOS, L. R. & SHER, A. (1997). Identification and characterization of protozoan products that trigger the synthesis of IL-12 by inflammatory macrophages. In *IL-12*. (ed. Aldorini, A.), *Chemical Immunology* **68**, 136–152. Basel: Karger.

GENTA, R. M. & WALZER, P. D. (1989). Strongyloidiasis. In *Parasitic Infections in the Compromised Host* (ed. Walzer, P. D. & Genta, R. M.), pp. 463–525. New York: Marcel Dekker.

GOOSSENS, B., OSAER, S., KORA, S., JAINTER, J., NDAO, M. & GEERTS, S. (1997). The interaction of *Trypanosoma congolense* and *Haemonchus contortus* in Djallonké sheep. *International Journal for Parasitology* **27**, 1579–1584.

GREENWOOD, B. M. (1974). Immunosuppression in malaria and trypanosomiasis. In *Parasites in the Immunized Host: Mechanisms of Survival*. Ciba Foundation Symposium No. 25 (new series), pp. 137–146. Amsterdam: Elsevier.

GREENWOOD, B. M. (1997). The epidemiology of malaria. *Annals of Tropical Medicine and Parasitology* **91**, 763–769.

GRENCIS, R. K. & ENTWISTLE, G. M. (1997). Production of an interferon-gamma homologue by an intestinal nematode: functionally significant or interesting artefact? *Parasitology* **115**, S101–S105.

GRENFELL, B. T. & DOBSON, A. P. (eds) (1995). *Ecology of Infectious Diseases in Natural Populations*. Cambridge: Cambridge University Press.

GUERRERO, O. M., CHINCHILLA, M. & ABRAHAMS, E. (1997). Increasing of *Toxoplasma gondii* (Coccidia Sarcocystidae) infections by *Trypanosoma lewisi* (Kinetoplastida, Trypanosomatidae) in white rats. *Revista de Biologia Tropical* **45**, 877–888.

HAMADIEN, M., BAKHIET, M. & HARRIS, R. A. (2000). Interferon-γ induces secretion of trypanosome lymphocyte triggering factor via tyrosine protein kinases. *Parasitology* **120**, 281–287.

HEATH, D. D. (1995). Immunology of *Echinococcus infections*. In *Echinococcus and Hydatid Disease* (ed. Thompson, R. C. A. & Lymbery, A. J.) pp. 183–200. Wallingford, Oxford: CAB International.

HELMBY, H., KULLBERG, M. & TROYE-BLOMBERG, M. (1998). Altered immune responses in mice with concomitant *Schistosoma mansoni* and *Plasmodium chabaudi* infections. *Infection and Immunity* **66**, 5167–5174.

HIGGINS-OPITZ, S. B., DETTMAN, C. D., DINGLE, C. E., ANDERSON, C. B. & BECKER, P. J. (1990). Intestinal parasites of conventionally maintained BALB/c mice and *Mastomys coucha* and the effects of a concomitant schistosome infection. *Laboratory Animal Science* **24**, 246–252.

HILLGARTH, N. & WINGFIELD, J. C. (1995). Testosterone and immunosuppression in vertebrates: implications for parasite-mediated sexual selection. In *Parasites and Pathogens. Effects on Host Hormones and Behavior* (ed. Beckage, N. E.), pp. 143–155. New York: International Thomson Publishing.

HORNOK, S., HEIJMANS, J. F., BEKESI, L., PEEK, H. W., DOBOS-KOVACS, M., DREN, C. N. & VARGA, I. (1998). Interaction of chicken anaemia virus and *Cryptosporidium baileyi* in experimentally infected chickens. *Veterinary Parasitology* **31**, 43–55.

HOUBA, V. (1988). Specific immunity: immunopathology and immunosuppression. In *Malaria: Principles and Practice of Malariology* (ed. Wernsdorfer, W. H. & McGregor, I.), pp. 621–637. Edinburgh: Churchill Livingstone.

HOVETTE, P., CAMARA, P., DONZEL, C. & RAPHENON, G. (1999). Paludisme et tuberculose pulmonaire: effet 'booster' du paludisme sur la tuberculose? *Presse Médicale* **28**, 398–399.

HSU, D. Y. M. & GEIMAN, Q. M. (1952). Synergistic effect of *Haemobartonella muris* on *Plasmodium berghei* in white rats. *American Journal of Tropical Medicine and Hygiene* **1**, 747–760.

HUDSON, K. M. & TERRY, R. J. (1979). Immunodepression and the course of infection of a chronic *Trypanosoma brucei* infection in mice. *Parasite Immunology* **1**, 317–326.

HUGHES, F. W. & TATUM, A. L. (1956). Effects of hypoxia and intercurrent infections on infections by *Plasmodium berghei* in rats. *Journal of Infectious Diseases* **98**, 38–43.

ISHAM, V. & MEDLEY, G. (eds) (1996). *Models for Infectious Human Diseases: Their Structure and Relation to Data*. Cambridge: Cambridge University Press.

JAHIEL, R. I., NUSSENZWEIG, R. S., VANDERBERG, J. & VILCEK, J. (1968a). Antimalarial effect of interferon inducers at different stages of development of *Plasmodium berghei* in the mouse. *Nature* **220**, 710–711.

JAHIEL, R. I., VILCEK, J., NUSSENZWEIG, R. S. & VANDERBERG, J. (1968b). Interferon inducers protect mice against *Plasmodium berghei* malaria. *Science* **161**, 802–804.

KAYE, P. (1999). Parasite derived immunoregulatory molecules. *Parasite Immunology* **21**, 595–596.

KLEIN, J. & HOŘEJŠÍ, V. (1997). *Immunology*, 2nd edn., pp. 291–326 Oxford: Blackwell Science.

KLOETZEL, K., CHIEFFI, P. P., FALEIROS, J. J. & FILHO, T. J. (1977). Mortality and other parameters of concomitant infections in albino mice; the *Schistosoma-Toxoplasma* model. *Tropical and Geographical Medicine* **29**, 407–410.

KLOETZEL, K., FALEIROS, J. J. & MENDES, S. R. (1971). Concurrent infection of white mice with *T. cruzi* and *S. mansoni. Transactions of the Royal Society of Tropical Medicine and Hygiene* **65**, 530–531.

KRETTLI, A. U. (1977). Exacerbation of experimental *Trypanosoma cruzi* infection in mice by concomitant malaria. *Journal of Protozoology* **24**, 514–518.

LACROIX, C., BRUN-PASCAUD, M., MASLO, C., CHAU, F., ROMAND, S. & DEROUIN, F. (1996). Co-infection of *Toxoplasma gondii* with other pathogens: pathogenicity and chemotherapy in animal models. In *Toxoplasma gondii* (ed. Gross, U. J.). *Current Topics in Immunology and Microbiology*, No. 219, pp. 223–233.

LIN, D. S. & BOWMAN, D. D. (1992). Macrophage functions in cats experimentally infected with feline immunodeficiency virus and *Toxoplasma gondii. Veterinary Immunology and Immunopathology* **33**, 69–78.

MA, X., ASTE-AMEZAGA, M., GRI, G., GEROSA, F. & TRINCHIERI, G. (1997). Immunomodulatory functions and molecular regulation of IL-12. In *IL-12*. (ed. Aldorini, A), *Chemical Immunology* **68**, 1–22. Basel: Karger.

MABEY, D. C. W., BROWN, A & GREENWOOD, A. M. (1987). *Plasmodium falciparum* malaria and *Salmonella* infections in Gambian children. *Journal of Infectious Diseases* **155**, 1319–1321.

McGREGOR, I. A. & BARR, M. (1962). Antibody response to tetanus toxoid inoculation in malarious and non-malarious Gambian children. *Transactions of the Royal Society of Tropical Medicine and Hygiene* **56**, 364–367.

MADWAR, M. A., EL TAHAWY, M. & STRICKLAND, G. T. (1989). The relationship between uncomplicated schistosomiasis and hepatitis B infection. *Transactions of the Royal Society of Tropical Medicine and Hygiene* **83**, 233–236.

MAITLAND, K., WILLIAMS, T. N. & NEWBOLD, C. I. (1997). *Plasmodium vivax* and *P. falciparum*: biological interactions and the possibility of cross-species immunity. *Parasitology Today* **13**, 227–231.

MALHOTRA, I., MUNGAI, P., WAMACHI, A., KIOKO, J., OUMA, J. H., KAZURA, J. W. & KING, C. L. (1999). Helminths and Bacillus-Calmette-Guerin-induced immunity in children sensitized in utero to filariasis and schistosomiasis. *Journal of Immunology* **162**, 6843–6848.

MANGOUD, A. M., RAMADAN, M. E., MORSY, T. A., AMIN, A. M. & MOSTAFA, S. M. (1998). The histopathological picture of the liver of hamsters experimentally infected with *Leishmania d. infantum* on top of *Schistosoma mansoni* infection. *Journal of the Egyptian Society of Parasitology* **28**, 101–117.

MANSOUR, N. S., YOUSSEF, F. G., MIKHAIL, E. M. & MOHAREB, E. W. (1997). Amebiasis in schistosomiasis endemic and non-endemic areas in Egypt. *Journal of the Egyptian Society of Parasitology* **27**, 617–628.

MARKELL, E. K., JOHN, D. T. & KROTOSKI, W. A. (1999). *Markell and Voge's Medical Parasitology*, 8th edn. Philadelphia: W. B. Saunders.

MARSHALL, A. J., BRUNET, L. R., VAN GESSEL, Y., ALCARAZ, A., BLISS, S. K., PEARCE, E. J. & DENKERS, E. Y. (1999). *Toxoplasma gondii* and *Schistosoma mansoni* synergise to promote hepatocyte dysfunction association with high levels of plasma TNF-α and early death in C57BL/6 mice. *Journal of Immunology* **163**, 2089–2097.

MILLOTT, S. M. & COX, F. E. G. (1985). Interactions between *Trypanosoma brucei* and *Babesia* spp. and *Plasmodium* spp. in mice. *Parasitology* **90**, 241–254.

MING, M., EWEN, M. E. & PEREIRA, M. E. (1995). Trypanosome invasion of mammalian cells requires activation of the TGF beta signalling pathway. *Cell* **82**, 287–296.

MODRIĆ, S. & MAYBERRY, L. F. (1994). Effect of *Plasmodium berghei* (Apicomplexa) on *Nippostrongylus brasiliensis* (Nematoda) infection in the mouse *Mus musculus*. *International Journal for Parasitology* **24**, 389–395.

MORSY, T. A., MANGOUD, A. M., RAMADAN, M. E., MOSTAFA, S. M. & EL-SHARKAWY, I. M. (1998). The histopathology of the intestine of hamsters infected with *Leishmania d. infantum* on top of pre-existing schistosomiasis mansoni. *Journal of the Egyptian Society of Parasitology* **28**, 347–354.

MOSMANN, T. R. & COFFMAN, R. L. (1989). Heterogeneity of cytokine secretion patterns and functions of helper T cells. *Advances in Immunology* **46**, 111–147.

OMER, F. M., KURTZHALS, J. A. L. & RILEY, E. M. (2000). Maintaining the immunological balance in parasitic infections: a role for TGF-β. *Parasitology Today* **16**, 18–23.

ONAH, D. N. & WAKELIN, D. (1999). Trypanosome-induced suppression of responses to *Trichinella spiralis* in vaccinated mice. *International Journal for Parasitology* **29**, 1017–1026.

PETERS, W. (1965). Competitive relationship between *Eperythrozoon coccoides* and *Plasmodium berghei* in the mouse. *Experimental Parasitology* **16**, 158–166.

PETNEY, T. N. & ANDREWS, R. H. (1998). Multiparasite communities. *International Journal for Parasitology* **28**, 377–393.

PHILLIPS, R. S. & WAKELIN, D. (1976). *Trichuris muris*: effect of concurrent infections with rodent piroplasma on immune expulsion from mice. *Experimental Parasitology* **39**, 95–100.

POULIN, R. (1998). *Evolutionary Ecology of Parasites*. London, Chapman and Hall.

PRITCHARD, D. I., HEWITT, C. & MOQBEL, R. (1997). The relationship between immunological responsiveness controlled by T-helper 2 lymphocytes and infections with parasitic helminths. *Parasitology* **115**, S33–S44.

PURNOMO, SOLIHIN, A, GÓMEZ-SALADIN, E. & BANGS, M. J. (1999). Rare quadruple malaria infection in Irian Jaya Indonesia. *Journal of Parasitology* **85**, 574–579.

RIFAAT, M. A., SALEM, S. A., AZAB, M. E., EL-RAZIK, I. A., SAFER, E. H., BESHIR, S. R. & EL-SHENNAWY, S. F. (1984). Experimental concomitant toxoplasma and malaria infection in rats. *Folia Parasitologica* **31**, 97–104.

ROBERTS-THOMSON, I. C., GROVE, D. I., STEVENS, D. P. & WARREN, K. S. (1976). Suppression of giardiasis during the intestinal phase of trichinosis in the mouse. *Gut* **17**, 953–958.

ROBERTSON, L. J., CROMPTON, D. W. T., SANJUR, D. & NESHEIM, M. C. (1992). Haemoglobin concentrations and concomitant infections of hookworm and *Trichuris trichiura* in Panamanian primary schoolchildren. *Transactions of the Royal Society of Tropical Medicine and Hygiene* **86**, 654–656.

RODRÍGUEZ, M., TERRAZAS, L. I., MÁRQUEZ, R. & BOJALIL, R. (1999). Susceptibility to *Trypanosoma cruzi* is modified by a previous non-related infection. *Parasite Immunology* **21**, 177–185.

ROMAGNANI, S. (1991). Human Th1 and Th2: doubt no more. *Immunology Today* **12**, 256–257.

ROMAGNANI, S. (ed.) (1996). *Th1 and Th2 Cells in Health and Disease. Chemical Immunology* **63**. Basel: Karger.

ROOTH, I. B. & BJÖRKMAN, A. (1992). Suppression of *Plasmodium falciparum* infections during concomitant measles or influenza but not during pertussis. *American Journal of Tropical Medicine and Hygiene* **47**, 675–681.

ROSE, M. E., WAKELIN, D. & HESKETH, P. (1994). Interactions between infections with *Eimeria* spp. and *Trichinella spiralis* in inbred mice. *Parasitology* **108**, 69–75.

ROUSSEAU, D., LE FICHOUX, Y., STIEN, X, SUFFIA, I., FERRUA, B. & KUBAR, J. (1997). Progression of visceral leishmaniasis due to *Leishmania infantum* in BALB/c mice is markedly showed by prior infection with *Trichinella spiralis*. *Infection and Immunity* **65**, 4978–4983.

SALAMAN, M. H. (1969). Immunodepression by viruses. *Antibiotics and Chemotherapy* **15**, 393–406.

SALAMAN, M. H., WEDDERBURN, N. & BRUCE-CHWATT, L. J. (1969). The immunodepressive effect of a murine plasmodium and its interaction with murine oncogenic viruses. *Journal of General Microbiology* **59**, 383–391.

SERGENT, E. (1937). La prémunition dans le paludisme. *Rivista di Malariologia* **14** (Suppl.), pp. 5–25.

SERGENT, E. & PONCET, A. (1957). Étude expérimentale de l'association chez le rat blanc de la spirochétose hispano-nord-africaine et du paludisme des rongeurs a *Plasmodium berghei*. *Annales de l'Institut Pasteur Algerie* **35**, 1–23.

SHEHU, K. & NOWELL, F. (1998). Cross-reactions between *Eimeria falciformis* and *Eimeria pragensis* in mice induced by trickle infections. *Parasitology* **117**, 457–465.

SHMEUL, Z, GOLENSER, J. & SPIRA D. T. (1975). Mutual influence of infection with *Plasmodium berghei* and *Trypanosoma lewisi* in rats. *Journal of Protozoology* **22**, Abstract 73a.

SILVA, J. S., BARRAL-NETTO, M. & REED, S. G. (1993). Aggravation of both *Trypenosoma cruzi* and murine leukaemia virus by concomitant infections. *American Journal of Tropical Medicine and Hygiene* **49**, 589–597.

SMALLEY, M. E. (1975). The nature of age immunity to *Plasmodium berghei* in the rat. *Parasitology* **71**, 337–347.

SMITH, T., FELGER, I., TANNER, M. & BECK, H.-P. (1999). Premunition in *Plasmodium falciparum* infection:

insights from the epidemiology of multiple infections. *Transactions of the Royal Society of Tropical Medicine and Hygiene* **93** (Suppl. 1), S59–S64.

SMITHERS, S. R., TERRY, R. J. & HOCKLEY, D. J. (1969). Host antigens in schistosomiasis. *Proceedings of the Royal Society B* **171**, 483–494.

STEPHENSON, L. S. (1987). *Impact of Helminth Infections of Human Nutrition*. London: Taylor and Francis.

STERNBERG J. M. (1998). Immunobiology of African trypanosomiasis. In *Immunology of Intracellular Parasitism* (ed. Liew, F. Y. & Cox, F. E. G.), pp. 186–199. *Chemical Immunology* **70**. Basel: Karger.

STEWART, G. L., REDDINGTON, J. J. & HAMILTON, A. M. (1980). *Eimeria nieschulzi* and *Trichinella spiralis* in the rat. *Experimental Parasitology* **50**, 115–122.

STEWART, G. R., BOUSSINESQ, M., COULSON, T., ELSON, L., NUTMAN, T. & BRADLEY, J. E. (1999). Onchocerciasis modulates the immune response to mycobacterial antigens. *Clinical and Experimental Immunology* **117**, 517–523.

STRAUSS, E. & LACET, C. M. C. (1986). Hepatite e esquistossomose mansónica. In *Hepatites Agudas e Crónicas*. São Paulo: Sarvier.

TANNER, M. & BAKER, J. R. (eds) (1999). *The Epidemiology of Multiple Plasmodium falciparum Infections. Transactions of the Royal Society of Tropical Medicine and Hygiene* **93** (Suppl. 1), S1–S68.

TAYLOR-ROBINSON, A. W. & PHILLIPS, R. S. (1998). Infective dose moderates the balance between Th$_1$- and Th$_2$-regulated immune responses during blood-stage malaria. *Scandinavian Journal of Immunology* **48**, 527–544.

TOFT, C. A., AESCHLIMANN, A. & BOLIS, L. (eds) (1991). *Parasite-Host Associations: Coexistence or Conflict?* Oxford: Oxford University Press.

URQUHART, G. M., MURRAY, M., MURRAY, F. W., JENNINGS, F. W. & BATE, E. (1973). Immunodepression in *Trypanosoma brucei* infections in rats and mice. *Transactions of the Royal Society of Tropical Medicine and Hygiene* **65**, 528–535.

VERINAUD, L., DA CRUZ-HÖFLING, M. A., SAKURADA, J. K., RANGEL, H. A., VASSALLO, J., WAKELIN, D., SEWELL, H. F. & CAMARGO, I. J. (1998). Immunodepression induced by *Trypanosoma cruzi* and mouse hepatitis virus 3 is associated with thymus apoptosis. *Clinical Diagnostic Laboratory Immunology* **5**, 186–191.

VIEIRA, L. Q., OLIVIERA, M. R., NEUMANN, E., NICOLL, J. R. & VIEIRA, E. C. (1998). Parasitic infections in germfree mice. *Brazilian Journal of Medical and Biological Research* **31**, 105–110.

VINAYAK, V. K. & CHOPRA, A. K. (1978). The interaction between *Entamoeba histolytica* and *Syphacia obvelata* infection in mice. *Annals of Tropical Medicine and Parasitology* **72**, 549–551.

VOLLER, A., GARNHAM, P. C. C. & TARGETT, G. A. T. (1966). Cross immunity in monkey malaria. *Journal of Tropical Medicine and Hygiene* **69**, 121–123.

WAKELIN, D. (1996). *Immunity to Parasites: How Parasitic Infections are Controlled*. Cambridge: Cambridge University Press.

WALSH, A. L., PHIRI, A. J., GRAHAM, S. M., MOLYNEUX, E. M. & MOLYNEUX, M. E. (2000). Bacteremia in febrile Malawian children: children and microbiologic features. *Pediatric Infectious Disease Journal* **19**, 312–318.

WEDDERBURN, N. (1970). Effect of concurrent malaria infection on development of virus-induced lymphoma in Balb/c mice. *Lancet* **2**, 1114–1116.

WEDDERBURN, N. (1974). Immunodepression produced by malarial infection in mice. In *Parasites in the Immunized Host. Mechanisms of Survival*. Ciba Foundation Symposium No. 25 (new series), pp. 123–135. Amsterdam: Elsevier.

WHITTLE, H. C., BROWN, J., MARSH, K., GREENWOOD, B. M., SEIDELIN, P., TIGHE, H. & WEDDERBURN, L. (1984). T-cell control of Epstein–Barr virus-infected B cells is lost during *P. falciparum* malaria. *Nature* **312**, 449–450.

WILLIAMSON, W. A. & GREENWOOD, B. M. (1978). Impairment of the immune response to vaccination after severe malaria. *Lancet* **1**, 1328–1329.

YOELI, M. (1956). Some aspects of concomitant infections of plasmodia and schistosomes. I. The effect of *Schistosoma mansoni* on the course of *Plasmodium berghei* in the field vole (*Microtus guentheri*). *American Journal of Tropical Medicine and Hygiene* **5**, 988–999.

YOELI, M., BECKER, Y. & BERNKOPF, H. (1995). [The effect of West Nile Virus on experimental malaria infection (*Plasmodium berghei*) in mice]. In Hebrew, *Harefuah, Jerusalem* **49**, 116–119.

YOSHIDA, A., MARUYAMA, H., YABU, Y., AMANO, T., KOBAYAKAWA, T. & OHTA, N. (1999). Immune responses against protozoal and nematodal infection in mice with underlying *Schistosoma mansoni* infection. *Parasitology International* **48**, 73–79.

Interactions involving intestinal nematodes of rodents: experimental and field studies

J. M. BEHNKE[1]*, A. BAJER[2], E. SINSKI[2] and D. WAKELIN[1]

[1] *School of Life and Environmental Sciences, University Park, University of Nottingham, Nottingham NG7 2RD, UK*
[2] *Department of Parasitology, Institute of Zoology, University of Warszawa, ul.Krakowskie Przedmiescie 26/28, 00-927 Warszawa, Poland*

SUMMARY

Multiple species infections with parasitic helminths, including nematodes, are common in wild rodent populations. In this paper we first define different types of associations and review experimental evidence for different categories of interactions. We conclude that whilst laboratory experiments have demonstrated unequivocally that both synergistic and antagonistic interactions involving nematodes exist, field work utilizing wild rodents has generally led to the conclusion that interactions between nematode species play no, or at most a minor, role in shaping helminth component communities. Nevertheless, we emphasize that analysis of interactions between parasites in laboratory systems has been fruitful, has made a fundamental contribution to our understanding of the mechanisms underlying host-protective intestinal immune responses, and has provided a rationale for studies on polyparasitism in human beings and domestic animals. Finally, we consider the practical implications for transmission of zoonotic diseases to human communities and to their domestic animals, and we identify the questions that merit research priority.

Key words: Rodents, nematodes, helminths, associations, co-occurrence, synergistic interactions, antagonistic interactions, field studies, experimental studies.

INTRODUCTION

All animals are exposed, on a daily basis, to a wide range of infectious organisms. Consequently, it is rare for them to be uninfected or even to carry just a single species of pathogen. Most animals can act as hosts for a parasite fauna comprising a range of species, and surveys have consistently shown that in nature the majority of individuals carry more than one species concurrently. Wild rodents in the UK show a mean helminth species richness of 2 or more (Lewis 1968 a, b; Montgomery & Montgomery, 1989; Behnke et al. 1999). This is a common finding in rodents elsewhere (Haukisalmi & Henttonen, 1993 a; Tenora & Stanek, 1995) which extends to many other mammals (Pence, Crum & Conti, 1983; Waid, Pence & Warren, 1985), birds (Bush & Holmes, 1986; Kennedy, Bush & Aho, 1986), and even to human beings. In regions where human helminth infections are endemic, multiple species infection (polyparasitism) is common, rather than exceptional (Buck et al. 1978; Keusch & Migasena 1982; Kvalsvig, 1988; Ashford, Craig & Oppenheimer 1992; Chunge et al. 1995).

The component community structure of helminths in wild rodents in any geographical region is dynamic and known to be influenced by both extrinsic (year, season, site) and intrinsic (host sex, age, reproductive status) factors (Haukisalmi, Henttonen & Tenora, 1988; Boggs et al. 1991; Abu-

Madi et al. 1998, 2000; Behnke et al. 1999). These, and other factors, combine to shape the component community structure in ways that vary from location to location (Mollhagan, 1978; Martin & Huffman, 1980; Montgomery & Montgomery, 1989; Abu-Madi et al. 1998). In general, extrinsic factors have a greater influence than intrinsic factors, but their precise individual and combined contributions have seldom been fully evaluated (e.g. seasonal influences may in part reflect seasonal changes in host reproductive hormone cycles and associated changes in immunocompetence). Perhaps the least understood factor in this context is the extent to which the component communities of helminths in rodents are shaped by interactions between the parasites themselves.

In this paper, focusing on rodent hosts, we review the evidence for interactions between parasites, in which at least one of the associates is an intestinal nematode. This paper therefore builds on the earlier review by Christensen et al. (1987) and, in the light of Petney & Andrews (1998), addresses the specific issue of whether intestinal nematodes predispose their hosts to infections by heterologous species. First, we define different types of associations between species and review experimental evidence for different categories of interactions. Then we evaluate the evidence for interactions between species in wild rodent populations and whether experimental studies have improved our understanding of the role of interactions in the component community structures of helminths in wild rodent populations. We emphasize that analysis of inter-

* Corresponding author: Tel: 0115 951 3208. Fax: 0115 951 321. E-mail: jerzy.behnke@nottingham.ac.uk

actions between parasites in laboratory systems has been a fruitful approach that has made a fundamental contribution to our understanding of the mechanisms underlying host-protective intestinal immune responses. Finally, we consider the practical implications for transmission of zoonotic diseases to human communities and to their domestic animals, and we identify the questions that merit research priority.

CATEGORIES OF INTERACTION

At its simplest level, co-occurrence of different species of parasites may reflect quite independent infection processes. For example, if infective stages of several parasites are aggregated in the environment, the individual hosts that frequent such sites will be more exposed to infection and as a result carry heavier mixed burdens than others. Infections are likely to accumulate with time and consequently, with increasing host age, parasite burdens may grow heavier and more diverse (mean helminth species richness is known to increase with host age in wood mice, Montgomery & Montgomery, 1989, Behnke *et al.* 1999). Variations in host behaviour, in particular in exposure-related activities, are believed to be the principal determinants of predisposition to infection in humans (Bundy & Blumenthal, 1990; Chan, Bundy & Kan, 1994) and probably in most wild mammals, including rodents (Barnard & Behnke, 1990). Associations may therefore be initiated in part by chance events combining with host behaviour, but their persistence (and therefore the likelihood of their detection) suggests that they may be maintained because the parasites concerned interact with one another, either directly or via the host.

Synergistic (positive) interactions

Nematodes may modify the internal environment within the host to make it more suitable for a second species, with the consequence that the second species stands a greater chance of survival in the presence of the first. This may arise because the first species weakens the host's defenses in order to facilitate its own survival, but thereby generates an opportunity for other species both to colonize the host and/or to survive longer than they otherwise would. This type of interaction was termed 'interactive protection from expulsion' by Mitchell (1979) and would be expected to lead to positive associations between species in wild host populations.

Antagonistic (negative) interactions

Parasites may also interact with each other antagonistically, leading to reduction in the abundance of the second species in animals harbouring the first species. Theoretically, antagonistic interactions between parasites may occur for a variety of reasons. Non-immunologically-mediated interactions may in-

clude competition for attachment sites or for limited nutritional resources within the host, the action of toxic metabolic products from one species against another, and physical displacement as a consequence of a bulky biomass (e.g. adult cestodes, Haukisalmi & Henttonen 1993b).

Immunologically-mediated antagonism may arise through specific or non-specific cross immunity. If two parasite species share antigens, a specific immune response elicited by one may be effective against the second. A negative interaction may also follow when the host response to the first species generates a hostile environment in which the second species cannot survive adequately. This type of interaction was termed 'interactive intestinal expulsion' by Mitchell (1979).

SYNERGISTIC INTERACTIONS INVOLVING NEMATODES

Non-immunologically-mediated interactions

There is little direct experimental evidence for non-immunological synergistic interactions in the intestine between rodent nematodes of the type that have been reported in other host/parasite combinations (Holmes, 1973). Although nematode infections profoundly alter the intestinal environment this usually has no effect on other species present concurrently or is deleterious. For example, no evidence was found for altered growth rates or effects on fecundity during concurrent infections with *Trichinella spiralis* and *Nippostrongylus brasiliensis* in mice, other than in the phase when mice responded to either parasite with intestinal inflammation (Kennedy, 1980). In rats exposed concurrently to *N. brasiliensis* and *Strongyloides ratti*, the establishment of both nematodes was much the same as in single infection controls (Nawa & Korenaga, 1983). Similarly, in three-way interactions between *Hymenolepis diminuta*, *Moniliformis moniliformis* and *N. brasiliensis* no evidence was found for synergistic interactions (Holland, 1987).

Species of nematodes found in concurrent infections often occur in different regions of the intestine, their distributions presumably determined by differences in the qualities of the niche occupied. This spatial separation may not favour non-immunological synergistic interactions and may in fact represent the consequences of antagonistic interactions (see below).

Interactions mediated through the host's immune system

Following a primary exposure the trichostrongyloid nematode *Heligmosomoides polygyrus* establishes long-lasting infections in mice. Experiments with the murine subspecies *H. p. bakeri* have shown that in most strains of laboratory mice infections last for

8–10 months (Robinson *et al.* 1989). The parasite survives because it immunomodulates the intestinal environment, secreting immunomodulatory factors (IMF) that interfere with the T cell- and cytokine-mediated regulation of the intestinal inflammatory response (Behnke, 1987; Monroy & Enriquez, 1992; Telford *et al.* 1998). In particular, *H. polygyrus* infection impairs the mucosal mast cell response (Dehlawi, Wakelin & Behnke 1987), although little is known about the molecular mechanisms involved. As might be expected, host protective immune responses against other species of intestinal nematodes, which depend on a functional mast cell response, are downregulated in concurrent infections, enabling the heterologous species to survive much longer than they otherwise would. For example, infections with *Trichinella spiralis*, which are normally terminated in 2–3 weeks, persist for more than 6 weeks when *H. polygyrus* is present (Behnke, Wakelin & Wilson 1978). Similarly infections with *Trichuris muris* are also prolonged (Jenkins & Behnke, 1977), enabling even moderately intense worm burdens to survive to patency (Behnke, Ali & Jenkins 1984). This non-specific interactive protection from expulsion even extends to tapeworms such as *Hymenolepis diminuta* and *H. citelli* (Hopkins, 1980; Alghali, Hagan & Robinson, 1985).

Experimental work involving concurrent infections with *H. polygyrus* and *T. spiralis* has shown that production of cytokines such as IL-9 and IL-10, which are required to elicit and sustain mucosal mastocytosis in response to infection with *T. spiralis*, is significantly inhibited in concurrent infections. The outcome is a less intense acute response, and a slower loss of *T. spiralis* (Behnke *et al.* 1993).

Trichuris muris is also believed to enhance its survival by host immunomodulation. During a primary infection Th_2-mediated host-protective immunity operates primarily against larval stages, being considerably less effective against adult worms (Grencis, 1997). If the host response is delayed and the worms develop to the pre-adult and adult stages, chronic infections result. These stages are believed to secrete IMFs that polarize the host response towards Th_1. Since such responses down-regulate Th_2 activity, the parasite burden is then not expelled. This dependency of immunity against *T. muris* upon functional Th_2 responses may well explain earlier observations that infections were prolonged in mice exposed to protozoan infections, where Th_1 responses predominate (Phillips, Selby & Wakelin, 1974; Phillips & Wakelin, 1976). Similar data were obtained in concurrent infections between *T. spiralis* and *Eimeria vermiformis* (Rose, Wakelin & Hesketh, 1994). Phillips & Wakelin (1976) used concurrent infections involving *Babesia microti*, a species that is prevalent in wild rodent populations and therefore may exert an influence on their nematode infections. Effects upon the protozoan infection were not

reported, but it is interesting to note that, more recently, in similar experiments involving *H. polygyrus*, it was demonstrated that the time-course of *Babesia microti* infections was not affected (Behnke, Sinski & Wakelin, 1999).

The possibility that nematode-induced immune modulation influencing concurrent infections may be more widespread than so far studied experimentally is reinforced by data showing that a number of species release excretory/secretory factors capable of down-regulating host T cell proliferation (Allen & MacDonald, 1998). It is also interesting to note that helminth-induced upregulation of Th_2 responses has been proposed as a dominant factor in the pathogenesis of AIDS in Africa, such responses, and the corresponding downregulation of Th_1 responses, making the host more susceptible to infection with HIV (Bentwich, Kalinkovich & Weisman, 1995).

ANTAGONISTIC INTERACTIONS INVOLVING
NEMATODES

Non-immunologically-mediated interactions

The niches occupied by adult nematodes in the intestine vary both longitudinally and radially. Longitudinally-distinct niches range from the oesophagus to the rectum, whilst radially-distinct niches include the lumen, the mucosal surface, the various layers of the mucosa itself and the serosa (Crompton, 1973). Niche selection begins at the establishment stage and is reflected in the choices made by incoming larvae (Holmes, 1961, 1973; Sommerville, 1963). Some have argued that the narrow location specificity of nematodes in the intestine is a reflection of past competitive interactions, leading to selective segregation and preferential selection of non-overlapping niches to minimize competition between species in order to achieve maximum fitness (Schad, 1963; Holmes, 1973). However, others have pointed out that since many niches in vertebrate hosts are unfilled (Price, 1980) niche segregation may be primarily driven by the need for parasites to aggregate for mating and reproduction (Rohde, 1979). Competitive interactions among helminth parasites in general have been thoroughly summarized by Christensen *et al.* (1987), and their long-term evolutionary consequences (i.e. extinction, and interactive or selective niche segregation) have been reviewed (Holmes, 1973), discussed (Rohde, 1979; Price, 1980) and modeled (Dobson, 1985).

Interactions mediated through the host's immune system

Non-specific inflammatory effector mechanisms. Intestinal inflammatory responses to infection with nematodes are controlled largely by Th lymphocytes through their release of cytokines that regulate the various effector arms (reviewed Behnke *et al.* 2000, Else & Finkelman, 1999). Whilst there is general agreement that host-protective intestinal responses require functional Th$_2$ cell activity, there is still controversy over what it is that actually causes worms to leave the host. One idea that has been current for many years is that the effectors released at the height of the acute response create an intestinal environment that is detrimental to worm survival. Early work using concurrent infections of *T. spiralis* and other parasites, suggested that the effector phase appeared to act non-specifically to bring about expulsion of heterologous species resident in the intestine at the time of the acute response to *T. spiralis*. Thus *T. muris* and *Nippostrongylus brasiliensis* were expelled prematurely when the host response to *T. spiralis* was maximal (Bruce & Wakelin, 1977; Kennedy, 1980) and *T. spiralis* itself was lost prematurely when the response to *N. brasiliensis* was maximal (Kennedy, 1980). Protozoan parasites such as *Giardia muris* (Roberts-Thomson *et al.* 1976), *Eimeria nieschulzi* (Stewart, Reddington & Hamilton, 1980) and *E. pragensis* (Rose *et al.* 1994) were also affected by inflammatory responses elicited by *T. spiralis*. Similarly, tapeworms such as *H. diminuta* and *Rodentolepis nana* were lost from *T. spiralis*-infected mice and rats (Behnke, Bland & Wakelin, 1977; Christie, Wakelin & Wilson, 1979; Ferretti *et al.* 1984). *Rodentolepis* (= *Hymenolepis*) *microstoma* suffered severe impairment to its growth but because of its relatively sheltered attachment position in the bile duct escaped expulsion from the intestine (Howard *et al.* 1978). Thus the picture which emerged from these studies was one in which an antigen-specific immune response elicited wide ranging and non-specific effector mechanisms, which removed other parasites out of the intestine at the time of, or immediately following, the acute response to the inducing species.

This picture was radically altered when Japanese workers published an elegant and informative set of experiments involving concurrent infections (reviewed Nawa *et al.* 1994). By exploiting nude (athymic) mice as a immunologically-inert host environment, they showed that injecting IL-4 (to induce a mastocytosis) into mice carrying both *N. brasiliensis* and *Strongyloides ratti*, resulted in the removal only of the latter species. This raised the concept of selectivity in the effectiveness of effector mechanisms, i.e. intestinal mucosal mastocytosis was essential for the elimination of *S. ratti*, but was a redundant effector in the case of *N. brasiliensis*. Instead, the latter succumbs to the infection-induced mucus response even though the products of activated mast cells are released during the course of infection (Nawa & Korenaga, 1983). Later studies (Khan *et al.* 1993) showed that mastocytosis was also a key effector against *Strongyloides venezuelensis*, suggesting a genus-specific susceptibility.

Effect of non-specifically-elicited effectors on chronic nematode species. Experiments involving concurrent infections have been very informative in helping to understand how chronic nematode species survive in their hosts. When mice carrying low intensity infections with *H. polygyrus* (i.e. levels which minimize the immunomodulatory effects) were challenged with *T. spiralis*, the adult *H. polygyrus* were lost in proportion to the dose of *T. spiralis* administered (Behnke, Cabaj & Wakelin 1992). Since it is known that there is a positive correlation between the dose of *T. spiralis* and the intensity of the mucosal response induced (Dehlawi & Wakelin, 1995) this result indicates that *H. polygyrus* is susceptible in a dose-dependent manner to the non-specific effectors of the mucosal response. Therefore, the survival strategy of *H. polygyrus* may depend more on preventing intestinal responses from occurring rather than on tolerating the hostile intestinal environment once elicited (Smith & Bryant, 1986).

In contrast to *H. polygyrus*, hookworms were not eliminated during the course of a concurrent infection with *T. spiralis* (Behnke, Rose & Little, 1994). The adults of neither *Necator americanus* nor *Ancylostoma ceylanicum* were lost when hamsters rejected superimposed infections with *T. spiralis*, indicating that in these species resilience to the non-specific components of inflammatory responses are key features of the host-parasite relationship. How hookworms manage to survive in experimental model systems (e.g. hamsters) or in their canine and human hosts is still largely unknown, but current interest centres on hookworm products that alter the environment locally where the mouth-parts embed in the mucosa (Hotez & Pritchard, 1995). The combination of firm, but not permanent attachment (hookworms move feeding stations at regular intervals, Kalkofen, 1970), bioactive chemicals and oxygen radical scavenging enzymes (Brophy & Pritchard, 1992) may endow hookworms with sufficient resilience to survive acute responses to other parasites.

Immunologically-specific interactions

Cross immunity. There is very considerable antigenic cross-reactivity between parasitic nematodes, a fact that bedevils immunodiagnosis. Such cross-reactivity may also extend to antigens involved in host-protective immune responses. There have been several records of cross-resistance between intestinal nematodes although an antigenic basis for the resistance has not always been demonstrated. One example where this has been shown concerns *Trichinella spiralis* and *Trichuris muris*, members of the same taxonomic group. Mice exposed to primary infection with *T. spiralis* or *T. muris* showed enhanced resistance to a subsequent challenge with the other species (Lee, Grencis & Wakelin, 1982) and this resistance could be induced both by active immunization with an antigen preparation of each species and by adoptive transfer of immune lymphocytes.

Enhancement of Th₂-based responsiveness. Just as the polarization of host immunity towards Th_1 responses may explain some examples of synergistic interactions (see above), the converse may provide a theoretical basis for antagonistic interactions. Experimental studies using *T. muris*, where the susceptibility of certain inbred mouse strains reflects the dominance of Th_1 responses, have shown that susceptibility can be overridden by concurrent infections with parasites that induce strong specific Th_2 responses. This has been demonstrated using *T. spiralis* and *T. muris* (Hermanek, Goyal & Wakelin, 1994) and *Schistosoma mansoni* and *T. muris* (Curry *et al.* 1995). In both cases the Th changes were reflected in altered antibody isotype and cytokine profiles.

DO QUANTITATIVE INTERACTIONS INVOLVING NEMATODES EXIST IN NATURAL RODENT COMMUNITIES?

Evidence from the field – helminth communities in wild rodents

In host populations where total parasite species richness is high and where several core (dominant) species each show high prevalence, polyparasitism is likely to be common even in the absence of any synergistic interactions; indeed this is well recognized to be the case. Among randomly-associating helminth species, purely on grounds of probability, we would expect a preponderance of positive associations in component communities characterized by many common (prevalence > 90%) species and negative associations when rare (prevalence < 10%) species predominate (Lotz & Font, 1994).

Kisielewska (1970b) was among the first to address the question of whether such co-occurrences include interactions between species. Her studies revealed that some species of nematodes showed antagonistic associations (*Capillaria murissylvatici* with *Mastophorus muris*, *Heligmosomum halli* [now = *H. mixtum*, Tenora & Meszaros, 1971] with *Heligmosomoides glareoli*) whilst others showed synergistic associations e.g. *H. mixtum* with *Syphacia obvelata* [now = *S. petrusewiczi*, Tenora & Meszaros, 1975], *S. petrusewiczi* with *C. murissylvatici*, *S. petrusewiczi* with the cestode *Catenotaenia pusilla* (probably *C. henttoneni*, See Haukisalmi & Tenora, 1993), and *H. mixtum* with *C. henttoneni*. However, Kisielewska (1970b) employed simple statistics in her analyses and re-examination of her results failed to substantiate the validity of these associations other than that involving *H. mixtum* and *C. pusilla* (Hobbs, 1980).

Employing considerably more powerful statistical tests, Haukisalmi & Henttonen (1993a) came to the conclusion that significant associations of helminths were rare in *Clethrionomys glareolus* in Finland. Of the 12 pairs of helminths infecting their bank vole populations only 4 combinations (*H. mixtum* with *Heligmosomoides glareoli*, *C. murissylvatici* with *M. muris*, *H. mixtum* with *Catenotaenia* sp., and *H. mixtum* with the cestode *Paranoplocephala gracilis*), all positive, were identified in presence/absence data and only three showed significant quantitative associations (negative – *H. mixtum* with *Catenotaenia* sp., *H. glareoli* with *P. gracilis* and positive – *C. murissylvatici* with *M. muris*). Only two of these combinations interacted consistently (*H. mixtum* with *Heligmosomoides glareoli*, *C. murissylvatici* with *M. muris*), the other associations being weak and unpredictable with respect to the intrinsic and extrinsic factors that also influenced worm burdens.

Collectively these studies, and others, suggest that whilst helminths co-occur commonly in nature, associations (presence/absence data and quantitative) are rare and inconsistent when detected, often dependent on a particular season, site, or occurring only in specific years. Mostly, they are generated through factors other than interactions between the species concerned. Montgomery & Montgomery (1990), who studied nine species of helminths in wood mice (*Apodemus sylvaticus*) in Northern Ireland, arrived at much the same conclusion.

Unequivocal demonstration that synergistic or antagonistic interactions between parasites occur in naturally infected animals is fraught with statistical and logistic difficulties. Co-occurrence, and moreover, associations of particular pairs of species, do not necessarily imply that interactions between them exist as species of parasites may aggregate independently in particular hosts for a variety of reasons. To give just two examples, parasite transmission stages may selectively infect hosts of a particular sex and age (e.g. because they forage more extensively, as males do in the breeding season) and infective stages may be aggregated in a particular microenvironment into which some hosts wander more often than others. Even quantitative associations (as reflected in statistically significant positive/negative correlations) may arise simply because hosts show patterns of behaviour that predispose them to infection with both of a pair of species of parasites.

Taking the field systems into the laboratory

The laboratory strain *H. polygyrus bakeri* is widely used as a model of chronic intestinal nematode infections (Monroy & Enriquez, 1992). However, the fact that it was originally isolated from an abnormal host and has been passaged through various mouse strains for almost half a century (Behnke, Keymer & Lewis, 1991), suggests that this parasite is probably now quite different from that found in wild animals in the field. In Europe, wood mice (*Apodemus sylvaticus*) carry a closely related species *H. polygyrus polygyrus* (Lewis 1968 a; Montgomery & Montgomery, 1988). This species has been isolated recently and experiments in wood mice have shown that infections last some 12 weeks or so, i.e. less than *H. p. bakeri* in laboratory mice (Gregory, Keymer & Clarke, 1990) but considerably longer than species such as *T. spiralis* and *N. brasiliensis*. When this wild isolate was given to laboratory mice virtually no larvae survived the initial week of the tissue stage of infection (Quinnell, Behnke & Keymer, 1991). Whilst, intuitively we might suppose *H. p. polygyrus* to employ a similar, albeit perhaps somewhat weaker, immunomodulatory strategy to *H. p. bakeri*, to our knowledge this has not been investigated.

In Europe, wild voles carry nematodes of the related genus *Heligmosomum* (Tenora & Meszaros, 1971; Meszaros, 1977; Tenora & Stanek, 1995). In Poland, common voles, *Microtus arvalis*, show a high prevalence of infection with *Heligmosomum costellatum* (Bajer *et al*. unpublished). We have established both a breeding colony of the voles and cultures of the nematode to test hypotheses concerning immunomodulation as a survival strategy in this host-parasite combination. Preliminary data suggest that there are synergistic interactions between *H. costellatum* and the intestinal protozoan *Cryptosporidium parvum*, a species that infects domestic animals and is capable of causing severe enteritis in humans (James, 1997; Bednarska, Bajer & Sinski, 1998). Common voles have proved to be competent reservoir hosts of *C. parvum* and individuals infected with adult *H. costellatum* have been found to excrete significantly more oocysts than worm free animals from day 17 post infection with *C. parvum* until the end of the observation period (day 42) (Bajer *et al*. 2000).

Field tests of laboratory-generated hypotheses

While earlier quantitative survey data on helminth parasites of wild wood mice and voles did not provide strong support for interactions between species, laboratory experiments with *H. p. bakeri* would suggest that synergistic interactions should be found between the related Heligmosominae (*H. p. polygyrus*, *H. costellatum* and *H. mixtum*) and other parasites. In a recent re-analysis of such data Behnke *et al*. (unpublished) found that whilst *H. p. polygyrus* was associated positively with other species in terms of categorical data (presence/absence), only one species, the cestode *C. pusilla*, showed a quantitative positive interaction, the number of *C. pusilla* increasing with increasing burdens of *H. p. polygyrus*. Perhaps surprisingly, after taking account of differences between sites, seasons and host intrinsic factors such as age and sex, there was little evidence of

positive interactions between *H. p. polygyrus* and *Syphacia stroma*, two species that occupy a similar intestinal site and represent the dominant helminths of wood mice in the U.K.

PRACTICAL IMPLICATIONS

Understanding the extent to which nematodes predispose their rodent hosts to infection by heterologous species (whether nematodes or other parasites) is an important consideration, because wild rodents can act as reservoir hosts for parasites that can be transmitted to humans (zoonoses) and domestic animals. It is known that some parasites interfere with their host's defense mechanisms in order to facilitate their own survival and many of these species also show seasonal cycles of abundance in wild host populations. It might be expected, therefore, that such species could predispose hosts to co-infection by other parasites and other pathogens of potential medical significance at particular times of the year. If this were to be in the summer, coinciding with times when there is greater human contact with wildlife habitats, the potential for transmission would be greatly increased.

Some studies have shown that the prevalence and abundance of intestinal parasites in wood mice peaks in late autumn, winter and early spring (Kisielewska, 1970a; Montgomery & Montgomery, 1988; Abu-Madi *et al*. 1998). These seasons may be times of hardship for wild rodents which, although not normally regarded as commensal with humans, may move into human habitations for shelter. Wood mice, for example, often enter houses in the UK during the winter, at precisely the time when they carry the heaviest infections with *H. polygyrus*.

WHERE FROM HERE?

Perhaps the most interesting concept to emerge from this review in respect of interactions between nematodes and other parasites, is that there is a significant discrepancy between the findings from laboratory experiments and the observations gathered from wild rodents in the field. Whilst field workers have generally concluded that interactions between nematodes and other parasites play a minimal role in shaping the structures of helminth component communities, laboratory experiments have unequivocally demonstrated that under precisely controlled laboratory conditions such interactions do exist and can be quite drastic in their effects on particular species.

The strongest effects detected in laboratory experiments depend on positive or negative influences on the host's immune response by one of the interacting species. Intestinal immune responses are transient phenomena that occur over a period of days, or weeks at most, since sustaining regular and pro-

longed bouts of intestinal inflammation is itself
counter-productive for the host (Behnke *et al.* 2000).
The failure to detect interactions among parasites of
wild hosts reflects the difficulty of sampling sufficient
animals with concurrent infections at a time when
intestinal responses are taking place, and thus
obtaining adequate data to generate a statistically
significant result. Equally, parasite burden data
derived from culled wild hosts, whether sampled at
regular intervals or just in a restricted period,
provide only snapshot information on current parasite burdens, some of which might or might not be
influenced by past exposure to either homologous or
heterologous infections. An additional factor is that
low helminth burdens may not generate host-
protective immune responses (Behnke & Wakelin,
1973 – or may do the opposite – Bancroft, Else &
Grencis, 1994), and the threshold for generating
protective responses may be reached at different
times, as worm burdens accumulate at different rates
in individual hosts. The average life-span of wood
mice and voles is only 2–3 months during the
summer season (Gliwicz, 1983; Flowerdew 1985) so
that there may be a limited opportunity for slowly
developing intestinal immunity to become manifest.

However, where one species interferes non-specifically with host immunocompetence and is long-
lived, as in the case of *H. polygyrus*, we might expect
the probability of observing positive interactions to
be higher. The lack of strong support for quantitative
synergistic interactions between the heligmoso-
matids and other species (Haukisalmi & Henttonen,
1993*b*) suggests that in nature, if they survive by
immunomodulation, the effects of their evasive
strategies are confined to the site of localization in
the gut. In contrast to laboratory experiments, in
which inocula of the order of 100–200 larvae have
frequently been used (Jenkins & Behnke, 1977;
Behnke *et al.* 1978), parasite burdens of just 1–10
adult heligmosomatid worms may be insufficient to
prevent the expression of immunity to other parasites
(Pritchard & Behnke, 1985), and hence too low to
generate positive associations.

Most studies of helminths in wild rodents rely
simply on presence/absence data or on total worm
burdens (numbers of parasites), and few have been
more ambitious in collecting data on individual
parasite biomass, parasite fecundity or location (See
Moore & Simberloff, 1990 for helminth communities
in quail). Since, in comparison to laboratory experiments, wild animals generally harbour low worm
burdens, immune responses are likely to be milder in
naturally infected hosts and in consequence their
effects on heterologous species more subtle. Nevertheless, subtle effects on intestinal distribution have
been observed. Haukisalmi & Henttonen (1993*b*)
noted that while common nematodes had some, but
inconsistent, effects on the intestinal distribution of
other helminths, rare parasites such as *Capillaria* sp.

and the cestode *P. gracilis*, but also the more common
Catenotaenia sp., were associated with minor but
significant displacement of *H. mixtum* from its usual
location.

As we hope to have demonstrated, the study of
interactions between nematode and other parasites
has far broader significance than in just rodent
helminthology. Human polyparasitism is a major
problem in regions where human GI nematodes are
common and the exacerbation of morbidity and
pathology in multiple species infections is an important consideration, both in human and veterinary
medicine. Under infection regimes which were not
too dissimilar to those animals might experience
under conventional husbandry, it has been shown
that intestinal nematodes of cattle can synergise to
increase host susceptibility to lungworms (Kloosterman, Ploeger & Frankena, 1990). In concurrent
infections with *Ostertagia leptoaspicularis* and other
bovine *Ostertagia* species establishment of both
species was greater than when each was given alone
(Al Saqur *et al.* 1984). Awareness of the specificity of
protective responses, and in the light of rodent
model systems, it was demonstrated that there was
little cross immunity between common bovine GI
nematodes (Adams, Anderson & Windon, 1989)
raising disappointment for those aiming to develop
polyvalent wide-spectrum vaccines (Emery & Wagland, 1991).

It is our view that there are still many exciting and
relevant aspects of this topic that require further
attention. The analysis of interactions between pairs
of parasites in laboratory experiments has been a
fruitful exercise and has not yet been exhausted.
Each species of parasite is uniquely adapted to
survival in its host and analysis of the diversity of
their interactions has contributed to the development
of ecological models of competitive interactions
between species as well as shedding light on
fundamental aspects of host protective immunity
against parasites. It is vitally important to know
whether or not laboratory-generated data on
immune-based interactions can be extrapolated to
natural populations of hosts, be these humans,
domestic or wild animals. This information that in
practice must come largely from field and laboratory
work on nematodes of rodents, is necessary to
provide a rational framework for informed decisions
about research into and implementation of appropriate preventative control measures.

ACKNOWLEDGEMENTS

We are grateful to the British Council, UK and the State
Committee for Scientific Research (KBN), Poland for
financial support for our field studies (KBN Grants UM
855 and UM 930). We thank our colleagues, collaborators,
postgraduate and undergraduate students who have contributed to, motivated and inspired our research activities
in this field.

REFERENCES

ABU-MADI, M. A., BEHNKE, J. M., LEWIS, J. W. & GILBERT, F. S. (1998). Descriptive epidemiology of *Heligmosomoides polygyrus* in *Apodemus sylvaticus* from three contrasting habitats in south-east England. *Journal of Helminthology* **72**, 93–100.

ABU-MADI, M. A., BEHNKE, J. M., LEWIS, J. W. & GILBERT, F. S. (2000). Seasonal and site specific variation in the component community structure of intestinal helminths in *Apodemus sylvaticus* from three contrasting habitats in south-east England. *Journal of Helminthology* **74**, 7–16.

ADAMS, D. B., ANDERSON, B. H. & WINDON, R. G. (1989). Cross immunity between *Haemonchus contortus* and *Trichostrongylus colubriformis* in sheep. *International Journal for Parasitology* **19**, 717–722.

ALGHALI, S. T. O., HAGAN, P. & ROBINSON, M. (1985). *Hymenolepis citelli* (Cestoda) and *Nematospiroides dubius* (Nematoda): interspecific interactions in mice. *Experimental Parasitology* **60**, 369–370.

ALLEN, J. E. & MacDONALD, A. S. (1998). Profound suppression of cellular proliferation mediated by the secretions of nematodes. *Parasite Immunology* **20**, 241–247.

AL SAQUR, I., ARMOUR, J., BAIRDEN, K., DUNN, A. M., JENNINGS, F. W. & MURRAY, M. (1984). Experimental studies on the interaction between infections of *Ostertagia leptospicularis* and other bovine *Ostertagia* species. *Zeitschrift für Parasitenkunde* **70**, 809–817.

ASHFORD, R. W., CRAIG, P. S. & OPPENHEIMER, S. J. (1992). Polyparasitism on the Kenyan coast. 1. Prevalence and association between parasitic infections. *Annals of Tropical Medicine and Parasitology* **86**, 671–679.

BAJER, A., BEHNKE, J. M., BEDNARSKA, M., KANICKA, M. & SINSKI, E. (2000). The common vole (*Microtus arvalis*) as a competent host for *Cryptosporidium parvum*. (EMOP VIII, Poznan, 10–14 September, 2000.) *Acta Parasitologica* **45**, 178.

BANCROFT, A. J., ELSE, K. J. & GRENCIS, R. K. (1994). Low-level infection with *Trichuris muris* significantly affects the polarization of the CD4 response. *European Journal of Immunology* **24**, 3113–3118.

BARNARD, C. J. & BEHNKE, J. M. (Eds) (1990). *Parasitism and Host Behaviour*. London, Taylor & Francis.

BEDNARSKA, M., BAJER, A. & SINSKI, E. (1998). Calves as a potential reservoir of *Cryptosporidium parvum* and *Giardia* spp. *Annals of Agricultural and Environmental Medicine* **5**, 135–138.

BEHNKE, J. M. (1987). Evasion of immunity by nematode parasites causing chronic infections. *Advances in Parasitology* **26**, 1–71.

BEHNKE, J. M., ALI, N. M. H. & JENKINS, S. N. (1984). Survival to patency of low level infections with *Trichuris muris* in mice concurrently infected with *Nematospiroides dubius*. *Annals of Tropical Medicine and Parasitology* **78**, 509–517.

BEHNKE, J. M., BLAND, P. W. & WAKELIN, D. (1977). The effect of the expulsion phase of *Trichinella spiralis* on *Hymenolepis diminuta* infection in mice. *Parasitology* **75**, 79–88.

BEHNKE, J. M., CABAJ, W. & WAKELIN, D. (1992). The susceptibility of adult *Heligmosomoides polygyrus* to intestinal inflammatory responses induced by heterologous infection. *International Journal for Parasitology* **22**, 75–86.

BEHNKE, J. M., KEYMER, A. E. & LEWIS, J. W. (1991). *Heligmosomoides polygyrus* or *Nematospiroides dubius*? *Parasitology Today* **7**, 177–179.

BEHNKE, J. M., LEWIS, J. W., MOHD ZAIN, S. N. & GILBERT, F. S. (1999). Helminth infections in *Apodemus sylvaticus* in southern England: interactive effects of host age, sex and year on the prevalence and abundance of infections. *Journal of Helminthology* **73**, 31–44.

BEHNKE, J. M., LOWE, A., MENGE, D., IRAQI, F. & WAKELIN, D. (2000). Mapping the genes for resistance to gastrointestinal nematodes. *Acta Parasitologica* **45**, 1–13.

BEHNKE, J. M., ROSE, R. & LITTLE, J. (1994). Resistance of the hookworms *Ancylostoma ceylanicum* and *Necator americanus* to intestinal inflammatory responses induced by heterologous infection. *International Journal for Parasitology* **24**, 425–431.

BEHNKE, J. M., SINSKI, E. & WAKELIN, D. (1999). Primary infections with *Babesia microti* are not prolonged by concurrent *Heligmosomoides polygyrus*. *Parasitology International* **48**, 183–187.

BEHNKE, J. M., WAHID, F. N., GRENCIS, R. K., ELSE, K. J., BENSMITH, A. W. & GOYAL, P. K. (1993). Immunological relationships during primary infection with *Heligmosomoides polygyrus* (*Nematospiroides dubius*): downregulation of specific cytokine secretion (IL-9 and IL-10) correlates with poor mastocytosis and chronic survival of adult worms. *Parasite Immunology* **15**, 415–421.

BEHNKE, J. M. & WAKELIN, D. (1973). The survival of *Trichuris muris* in wild populations of its natural hosts *Parasitology* **67**, 157–164.

BEHNKE, J. M., WAKELIN, D. & WILSON, M. M. (1978). *Trichinella spiralis*: Delayed rejection in mice concurrently infected with *Nematospiroides dubius*. *Experimental Parasitology* **46**, 121–130.

BENTWICH, Z., KALINKOVICH, A. & WEISMAN, Z. (1995). Immune activation is a dominant factor in the pathogenesis of African AIDS. *Immunology Today* **16**, 187–191.

BOGGS, J. F., McMURRAY, S. T., LESLIE, D. M. JR., ENGLE, D. M. & LOCHMILLER, R. L. (1991). Influence of habitat modification on the community of gastrointestinal helminths of cotton rats. *Journal of Wildlife Diseases* **27**, 584–593.

BROPHY, P. M. & PRITCHARD, D. I. (1992). Immunity to helminths: ready to tip the biochemical balance? *Parasitology Today* **8**, 419–422.

BRUCE, R. G. & WAKELIN, D. (1977). Immunological interaction between *Trichinella spiralis* and *Trichuris muris* in the intestine of the mouse. *Parasitology* **74**, 163–173.

BUCK, A. A., ANDERSON, R. I., MACRAE, A. A. & FAIN, A. (1978). Epidemiology of poly-parasitism. I. Occurrence, frequency and distribution of multiple infections in rural communities on Chad, Peru, Afganistan and Zaire. *Tropical Medicine and Parasitology* **29**, 61–70.

BUNDY, D. A. P. & BLUMENTHAL, U. J. (1990). Human behaviour and the epidemiology of helminth infections: the role of behaviour in exposure to infection. In *Parasitism and Host Behaviour* (ed. Barnard, C. J. & Behnke, J. M.), pp. 264–289. London, Taylor & Francis.

BUSH, A. O. & HOLMES, J. C. (1986). Intestinal helminths of lesser scaup ducks: an interactive community. *Canadian Journal of Zoology* **64**, 142–152.

CHAN, L., BUNDY, D. A. P. & KAN, S. P. (1994). Genetic relatedness as a determinant of predispostition to *Ascaris lumbricoides* and *Trichuris trichiura* infection. *Parasitology* **108**, 77–80.

CHRISTENSEN, N. O., NANSEN, P., FAGBEMI, B. O. & MONRAD, J. (1987). Heterologous antagonistic interactions between helminths and between helminths and protozoans in concurrent experimental infection of mammalian hosts. *Parasitology Research* **73**, 387–410.

CHRISTIE, P. R., WAKELIN, D. & WILSON, M. M. (1979). The effect of the expulsion phase of *Trichinella spiralis* on *Hymenolepis diminuta* infection in rats. *Parasitology* **78**, 323–330.

CHUNGE, R. N., KARUMBA, N., OUMA, J. H., THIONGO, F. W., STURROCK, R. F. & BUTTERWORTH, A. E. (1995). Polyparasitism in two rural communities with endemic *Schistosoma mansoni* infection in Machakos District, Kenya. *Journal of Tropical Medicine and Hygiene* **98**, 440–444.

CROMPTON, D. W. T. (1973). Sites occupied by some parasitic helminths in the alimentary tract of vertebrates. *Biological Reviews* **48**, 27–83.

CURRY, A. J., ELSE, K. J., JONES, F., BANCROFT, A., GRENCIS, R. K. & DUNNE, D. W. (1995). Evidence that cytokine-mediated immune interactions induced by *Schistosoma mansoni* alter disease outcome in mice concurrently infected with *Trichuris muris*. *Journal of Experimental Medicine* **181**, 769–774.

DEHLAWI, M. S. & WAKELIN, D. (1995). Dose dependency of mucosal mast cell responses in mice infected with *Trichinella spiralis*. *Research and Reviews in Parasitology* **55**, 21–24.

DEHLAWI, M. S., WAKELIN, D. & BEHNKE, J. M. (1987). Suppression of mucosal mastocytosis by infection with the intestinal nematode *Nematospiroides dubius*. *Parasite Immunology* **9**, 187–194.

DOBSON, A. P. (1985). The population dynamics of competition between parasites. *Parasitology* **91**, 317–347.

ELSE, K. J. & FINKELMAN, F. D. (1999). Intestinal nematode parasites, cytokines and effector mechanisms. *International Journal for Parasitology* **28**, 1145–1158.

EMERY, D. L. & WAGLAND, B. M. (1991). Vaccines against gastrointestinal nematode parasites of ruminants. *Parasitology Today* **7**, 347–349.

FERRETTI, G., GABRIELE, F., PALMAS, C. & WAKELIN, D. (1984). Interactions between *Trichinella spiralis* and *Hymenolepis nana* in the intestine of the mouse. *International Journal for Parasitology* **14**, 29–33.

FLOWERDEW, J. R. (1985). The population dynamics of wood mice and yellow-necked mice. In *The Ecology of Woodland Rodents, Bank Voles and Wood Mice* (eds. Flowerdew, J. R., Gurnell, J. & Gipps, J. H. W.), pp. 315–338 Symposia of the Zoological Society of London **55**.

GLIWICZ, J. (1983). Survival and life span. In *Ecology Of The Bank Vole* (ed. Petrusewicz, K) *Acta Theriologica* **28** (suppl 1), 161–172.

GREGORY, R. D., KEYMER, A. E. & CLARKE, J. R. (1990). Genetics, sex and exposure: the ecology of *Heligmosomoides polygyrus* (Nematoda) in the wood mouse. *Journal of Animal Ecology* **59**, 363–378.

GRENCIS, R. K. (1997). Th2-mediated host protective immunity in intestinal nematode infections. *Philosophical Transactions of the Royal Society of London B* **352**, 1377–1384.

HAUKISALMI, V. & HENTTONEN, H. (1993a). Coexistence in helminths of the bank vole *Clethrionomys glareolus*. I. Patterns of co-occurrence. *Journal of Animal Ecology* **62**, 221–229.

HAUKISALMI, V. & HENTTONEN, H. (1993b). Coexistence in helminths of the bank vole *Clethrionomys glareolus*. II. Intestinal distribution and interspecific interactions. *Journal of Animal Ecology* **62**, 230–238.

HAUKISALMI, V., HENTTONEN, H. & TENORA, F. (1988). Population dynamics of common and rare helminths in cyclic vole populations. *Journal of Animal Ecology* **57**, 807–825.

HAUKISALMI, V. & TENORA, F. (1993). *Catenotaenia henttoneni* sp. n. (Cestoda: Catenotaeniidae), a parasite of voles *Clethrionomys glareolus* and *C. rutilus* (Rodentia). *Folia Parasitologica* **40**, 29–33.

HERMANEK, J., GOYAL, P. K. & WAKELIN, D. (1994). Lymphocyte, antibody and cytokine responses during concurrent infections between helminths that selectively promote T-helper-1 or T-helper-2 activity. *Parasite Immunology* **16**, 111–117.

HOBBS, R. P. (1980). Interspecific interactions among gastrointestinal helminths in pikas of North America. *American Midland Naturalist* **103**, 15–25.

HOLLAND, C. (1987). Interspecific effects between *Moniliformis* (Acanthocephala), *Hymenolepis* (Cestoda) and *Nippostrongylus* (Nematoda) in the laboratory rat. *Parasitology* **94**, 567–581.

HOLMES, J. C. (1961). Effects of concurrent infections on *Hymenolepis diminuta* (Cestoda) and *Moniliformis dubius* (Acantheocephala). I. General effects and comparison with crowding. *Journal of Parasitology* **47**, 209–216.

HOLMES, J. C. (1973). Site selection by parasitic helminths: interspecific interactions, site segregation and their importance to the development of the helminth communities. *Canadian Journal of Zoology* **51**, 333–347.

HOPKINS, C. A. (1980). Immunity and *Hymenolepis diminuta*. In *Biology of the Tapeworm* Hymenolepis diminuta (ed. Arai, H. P.), pp. 551–614. New York, Academic Press.

HOTEZ, P. J. & PRITCHARD, D. I. (1995). Hookworm infection. *Scientific American* **June 1995**, 42–48.

HOWARD, R. J., CHRISTIE, D., WAKELIN, D., WILSON, M. M. & BEHNKE, J. M. (1978). The effect of concurrent infection with *Trichinella spiralis* on *Hymenolepis microstoma* in mice. *Parasitology* **77**, 273–279.

JAMES, S. L. (1997). Emerging parasitic infections. *FEMS Immunology and Medical Microbiology* **18**, 313–317.

JENKINS, S. N. & BEHNKE, J. M. (1977). Impairment of primary expulsion of *Trichuris muris* in mice concurrently infected with *Nematospiroides dubius*. *Parasitology* **75**, 71–78.

KALKOFEN, U. P. (1970). Attachment and feeding behaviour of *Ancylostoma caninum*. *Zeitschirft für Parasitenkunde* **33**, 339–354.

KENNEDY, C. R, BUSH, A. O. & AHO, J. M. (1986). Patterns in helminth communities: why are birds and fish different? *Parasitology* **93**, 205–215.

KENNEDY, M. W. (1980). Immunologically mediated, non-specific interactions between the intestinal phases of *Trichinella spiralis* and *Nippostrongylus brasiliensis* in the mouse. *Parasitology* **80**, 61–72.

KEUSCH, G. T. & MIGASENA, P. (1982). Biological implications of polyparasitism. *Reviews of Infectious Diseases* **4**, 880–882.

KHAN, A. I., HORII, Y., TIURUA, R., SATO, Y. & NAWA, Y. (1993). Mucosal mast cells and the expulsive mechanisms of mice against *Strongyloides venezuelensis*. *International Journal for Parasitology* **23**, 551–555.

KISIELEWSKA, K. (1970*a*). Ecological organization of intestinal helminth groupings in *Clethrionomys glareolus* (Schreb.) (Rodentia). 1. Structure and seasonal dynamics of helminth groupings in a host population in the Białowieża National Park. *Acta Parasitologica Polonica* **18**, 121–147.

KISIELEWSKA, K. (1970*b*). Ecological organization of intestinal helminth groupings in *Clethrionomys glareolus* (Schreb.) (Rodentia). V. Some questions concerning helminth groupings in the host individuals. *Acta Parasitologica Polonica* **17**, 197–208.

KLOOSTERMAN, A., PLOEGER, H. W. & FRANKENA, K. (1990). Increased establishment of lungworms after exposure to a combined infection of *Ostertagia ostertagi* and *Cooperia oncophora*. *Veterinary Parasitology* **36**, 117–122.

KVALSVIG, J. D. (1988). The effects of parasitic infection on cognitive performance. *Parasitology Today* **4**, 206–208.

LEE, T. D. G., GRENCIS, R. K. & WAKELIN, D. (1982). Specific cross-immunity between *Trichinella spiralis* and *Trichuris muris*: immunization with heterologous infections and antigens and transfer of immunity with heterologous immune mesenteric lymph node cells. *Parasitology* **84**, 381–389.

LEWIS, J. W. (1968*a*). Studies on the helminth parasites of the long-tailed field mouse, *Apodemus sylvaticus sylvaticus* from Wales. *Journal of Zoology, London* **154**, 287–312.

LEWIS, J. W. (1968*b*). Studies on the helminth parasites of voles and shrews from Wales. *Journal of Zoology, London* **154**, 313–331.

LOTZ, J. M. & FONT, W. F. (1994). Excess positive associations in communities of intestinal helminths of bats: a refined null hypothesis and a test of the facilitation hypothesis. *Journal of Parasitology* **80**, 398–413.

MARTIN, J. L. & HUFFMAN, D. G. (1980). An analysis of the community and population dynamics of the helminths of *Sigmodon hispidus* (Rodentia: Criceditae) from three central Texas vegetational regions. *Proceedings of the Helminthological Society of Washington* **47**, 247–255.

MESZAROS, F. (1977). Parasitic nematodes of *Microtus arvalis* (Rodentia) in Hungary. *Parasitologica Hungarica* **10**, 67–83.

MITCHELL, G. F. (1979). Effector cells, molecules and mechanisms in host-protective immunity to parasites. *Immunology* **38**, 209–223.

MOLLHAGEN, T. (1978). Habitat influence on helminth parasitism of the cotton rat in western Texas, with remarks on some of the parasites. *The Southwestern Naturalist* **23**, 401–408.

MONROY, F. G. & ENRIQUEZ, F. J. (1992). *Heligmosomoides polygyrus*: a model for chronic gastrointestinal helminthiasis. *Parasitology Today* **8**, 49–54.

MONTGOMERY, S. S. J. & MONTGOMERY, W. I. (1988). Cyclic and non-cyclic dynamics in populations of the helminth parasites of wood mice *Apodemus sylvaticus*. *Journal of Helminthology* **62**, 78–90.

MONTGOMERY, S. S. J. & MONTGOMERY, W. I. (1989). Spatial and temporal variation in the infracommunity structure of helminths of *Apodemus sylvaticus* (Rodentia: Muridae). *Parasitology* **98**, 145–150.

MONTGOMERY, S. S. J. & MONTGOMERY, W. I. (1990). Structure, stability and species interactions in helminth communities of wood mice, *Apodemus sylvaticus*. *International Journal for Parasitology* **20**, 225–242.

MOORE, J. & SIMBERLOFF, D. (1990). Gastrointestinal helminth communities of bobwhite quail. *Ecology* **71**, 344–359.

NAWA, Y., ISHIKAWA, N., TSUCHIYA, K., HORII, Y., ABE, T., KHAN, A. I., BING-SHI, ITOH, H., IDE, H. & UCHIYAMA, F. (1994). Selective effector mechanisms for the expulsion of intestinal helminths. *Parasite Immunology* **16**, 333–338

NAWA, Y. & KORENAGA, M. (1983). Mast and goblet cell responses in the small intestine of rats concurrently infected with *Nippostrongylus brasiliensis* and *Strongyloides ratti*. *Journal of Parasitology* **69**, 1168–1170.

PENCE, D. B., CRUM, J. M. & CONTI, J. A. (1983). Ecological analyses of helminth populations in the black bear, *Ursus americanus*, from North America. *Journal of Parasitology* **69**, 933–950.

PETNEY, T. N. & ANDREWS, R. H. (1998). Multiparasite communities in animals and humans: frequency, structure and pathogenic significance. *International Journal for Parasitology* **28**, 377–393.

PHILLIPS, R. S., SELBY, G. R. & WAKELIN, D. (1974). The effect of *Plasmodium berghei* and *Trypanosoma brucei* infections on the immune expulsion of the nematode *Trichuris muris* from mice. *International Journal for Parasitology* **4**, 409–415.

PHILLIPS, R. S. & WAKELIN, D. (1976). *Trichuris muris*: effect of concurrent infections with rodent piroplasms on immune expulsion from mice. *Experimental Parasitology* **39**, 95–100.

PRICE, P. W. (1980). Evolutionary biology of parasites. *Monographs in Population Biology* **15**, Princeton University Press, Princeton, New Jersey, USA.

PRITCHARD, D. I. & BEHNKE, J. M. (1985). The suppression of homologous immunity by soluble adult antigens of *Nematospiroides dubius*. *Journal of Helminthology* **59**, 251–256.

QUINNELL, R. J., BEHNKE, J. M. & KEYMER, A. E. (1991). Host specificity of and cross-immunity between two strains of *Heligmosomoides polygyrus*. *Parasitology* **102**, 419–427.

ROBERTS-THOMSON, I. C., GROVE, D. I., STEVENS, D. P. & WARREN, K. S. (1976). Suppression of giardiasis during the intestinal phase of trichinosis in the mouse. *Gut* **17**, 953–958.

ROBINSON, M., WAHID, F. N., BEHNKE, J. M. & GILBERT, F. S. (1989). Immunological relationships during primary infection with *Heligmosomoides polygyrus* (*Nematospiroides dubius*): dose-dependent expulsion of adult worms. *Parasitology* **98**, 115–124.

ROHDE, K. (1979). A critical evaluation of intrinsic and extrinsic factors responsible for restriction in parasites. *The American Naturalist* **114**, 648–671.

ROSE, M. E., WAKELIN, D. & HESKETH, P. (1994). Interactions between infections with *Eimeria* spp. and *Trichinella spiralis* in inbred mice. *Parasitology* **108**, 69–75.

SCHAD, G. A. (1963). Niche diversification in a parasitic species flock. *Nature, London* **198**, 404–406.

SMITH, N. C. & BRYANT, C. (1986). The role of host generated free radicals in helminth infections: *Nippostrongylus brasiliensis* and *Nematospiroides dubius* compared. *International Journal for Parasitology* **16**, 617–622.

SOMMERVILLE, R. I. (1963). Distribution of some parasitic nematodes in the alimentary tract of sheep, cattle and rabbits. *Journal of Parasitology* **49**, 593–599.

STEWART, G. L., REDDINGTON, J. J. & HAMILTON, A. M. (1980). *Eimeria nieschulzi* and *Trichinella spiralis* in the rat. *Experimental Parasitology* **50**, 115–122.

TELFORD, G., WHEELER, D. J., APPLEBY, P., BOWEN, J. G. & PRITCHARD, D. I. (1998). *Heligmosomoides polygyrus* immunomodulatory factor (IMF) targets T-lymphocytes. *Parasite Immunology* **20**, 601–611.

TENORA, F. & MESZAROS, F. (1971). Nematodes of the genus *Heligmosomum* Railliet et Henry, 1909, sensu Durette-Desset, 1968, parasitizing rodents in Europe. *Acta Zoologica Academiae Scientiarum Hungaricae* **17**, 397–407.

TENORA, F. & MESZAROS, F. (1975). Nematodes of the genus *Syphacia*, Seurat, 1916 (Nematoda) – parasites of rodents (Rodentia) in Czechoslovakia and Hungary. *Acta Universitatis Agriculturae, Brno* **23**, 537–554.

TENORA, F. & STANEK, M. (1995). Changes of the helminthofauna in several muridae and Arvicolidae at Lednice in Moravia. II. Ecology. *Acta Universitatis Agriculturae et Silviculturae Mendelianae Brunensis* **43**, 57–65.

WAID, D. D., PENCE, D. B. & WARREN, R. J. (1985). Effects of season and physical condition on the gastrointestinal community of white-tailed deer from Texas Edwards Plateau. *Journal of Wildlife Diseases* **21**, 264–273.

Heterologous immunity revisited

I. A. CLARK*

Division of Biochemistry and Molecular Biology, School of Life Sciences, Australian National University, Canberra, Australia ACT 0200

SUMMARY

Heterologous immunity, or protection by one invading organism against another across phylogenetic divides, has been recognised for decades. It was initially thought to operate largely through enhancement of phagocytosis, but this explanation became untenable when it was realised it worked extremely well against intraerythrocytic protozoa and killed them while they were free in the circulation. Clearly a soluble mediator was called for. This review summarises the logic that arose from this observation, which led to a wider appreciation of the roles of pro-inflammatory cytokines, and then nitric oxide, in the host's response against invaders, as well as the ability of these mediators to harm the host itself if they are generated too enthusiastically. This has led to a discernable pattern across heterologous immunity as a whole, and its lessons influence a range of areas, including vaccine development.

Key words: BCG, Babesia, malaria, TNF, nitric oxide, disease pathogenesis, vaccines.

INTRODUCTION

Infectious disease researchers tend to think of 'their' disease in isolation, but in real life diseases happen together. The study of the heterologous immunity that occurs between such concomitant infections can have outcomes of immediate value. But more importantly, the basic biological concepts unearthed along the way have yielded secrets that have proved to have broad applicability. This review summarises where the study of the immunity observed when two unrelated infectious agents encounter each other in the one host has led our group, and thereby allowed others to be influenced by the literature that has developed.

These days our laboratory investigates the pathophysiology of the syndrome seen in severe falciparum malaria in children, as described by several groups working in Africa (Taylor, Borgstein & Molyneux, 1993; March *et al.* 1996), which typically includes metabolic acidosis and associated respiratory distress, hypoglycaemia, seizures, coma and cerebral oedema. I entered this area about 20 years ago, with no apparently relevant credentials, by making the then unlikely suggestion that excessive systemic production of pro-inflammatory cytokines, such as TNF (tumour necrosis factor) plays a key role in human malarial disease, gram-negative bacterial infection and the Jarisch-Herxheimer reaction (Clark *et al.* 1981). As its name implies, TNF was previously known only as a mediator that killed tumour cells, and until our interest in it had been worked on only by the Sloan Kettering group who named it. Cytokine-induced nitric oxide generated by inducible nitric oxide synthase (iNOS) was subse-

quently added to this model (Clark, Rockett & Cowden, 1992*b*). The harmful systemic effects of excess iNOS are now well recognised, with an ample literature (Ruetten *et al.* 1996; Schwartz *et al.* 1997; Numata *et al.* 1998) documenting the protective effect against illness of specific iNOS inhibitors in circumstances where pro-inflammatory cytokine levels are increased.

These concepts now drive much current work on malarial disease pathogenesis (Burgner *et al.* 1998; Kun *et al.* 1998; Knight *et al.* 1999; McGuire *et al.* 1999), and have paved the way to similar investigations in many other infectious diseases. While this work was effectively a revival of Brian Maegraith's insights of the 1940s (Maegraith, 1948) on malarial disease being a systemic inflammation, they actually arose, not from studying patients, but from trying to understand the basic nature of heterologous immunity between different parasites in mouse models. This review is about how experiments on heterologous immunity led us to the concept of involvement of the same pathways, involving the same mediators, in both host protection and host illness. This cast fresh light, from an unexpected quarter, on the nature of malarial disease, and our malaria experiments have in turn provided information that has fed back into understanding heterologous immunity, as well as giving a philosophical background into vaccine development.

BABESIA AND MALARIA

My interest in heterologous immunity, or immunity between phylogenetically unrelated organisms, began 25 years ago with Frank Cox pointing out to me the implications of the observations he had made a few years earlier that certain mutually cross-pro-

* Tel: + 61 2 6249 4363. Fax: + 61 2 6249 0313. E-mail: ian.clark@anu.edu.au

Parasitology (2001), **122**, S51–S59. Printed in the United Kingdom © 2001 Cambridge University Press

tecting malaria parasites (*Plasmodium chabaudi* and *P. vinckei*) also cross-protected with *Babesia microti* and *B. rodhaini*, but not with *P. berghei* or *P. yoelii* (Cox, 1970). Cross-reacting antibodies could not account for these observations, which went against all the rules of the specific antibody-dependent immunity thought, in those days, to control these infections. This effect was most graphically illustrated some years later by demonstrating the elimination of high densities of *P. vinckei*, a normally lethal malaria parasite, when it was present during the crisis phase of infections with the non-fatal organism, *B. microti* (Cox, 1978).

We were then trying to understand what killed *B. microti* and *P. chabaudi* parasites inside circulating red cells during the resolution of these infections (Clark *et al.* 1975), where antibody could not reach them. The manner of death of these parasites was in keeping with the 'crisis forms' described in monkeys in the 1930s (Taliaferro & Cannon, 1936). Cox's results encouraged us to see just how phylogenetically different an organism could be from *Plasmodium* or *Babesia* and still act in this heterologous protective fashion against these haemoprotozoa. On the hunch of a colleague, Jean-Louis Virelizier, that a newly-described type of interferon, later termed interferon-γ, might be involved (Salvin *et al.* 1975), we infected mice i.p. or i.v. with live BCG (the Bacillus Calmette-Guérin strain of *Mycobacterium tuberculosis*) several weeks before infecting them with haemoprotozoa. This was dramatically protective, particularly against *B. microti*, but also against malaria (Clark, Allison & Cox, 1976). Most revealingly, the parasites again died in circulating red cells, within 12 hours of being injected, and independently of antibody. The immunity was as strong and durable as that seen after recovery from a primary infection (Clark *et al.* 1977*b*). Killed *Corynebacterium parvum* had the same effects (Clark, Cox & Allison, 1977*a*). Again, protection against blood forms of *B. microti* and *B. rodhaini* was absolute, with no parasites being seen on smears, then or after repeated challenge.

BACTERIA, RICKETTSIAS AND INTRA-MACROPHAGE PROTOZOA

These results made us realise that we were dealing with a much wider question than we had originally thought, in that both BCG and *C. parvum* had been long-recognised to protect against tumours (Old, Clarke & Benacerraf, 1959; Halpern *et al.* 1966), bacteria (Dubos & Schaedler, 1957; Howard *et al.* 1959; Collins & Scott, 1974) and macrophage-dwelling protozoa (Swartzberg, Krahenbuhl & Remington, 1975; Smrkovski & Larson, 1977). These parallels continued to hold as we widened our range of macrophage-activating agents that gave this protection to include *Brucella abortus* Strain 19

(Herod, Clark & Allison, 1978), an extract of *Coxiella burnetii* (Clark, 1979*b*), and live *Salmonella enteriditis* and *Listeria monocytogenes* (Clark, 1979*a*). We also became aware that other researchers, as puzzled as we were, had been reporting that these infectious agents cross-protected against each other (Nyka, 1957; Jespersen, 1976; Zinkernagel, 1976). Such unexplained cross-protections could be as strong as the real thing – in fact, in our hands an extract of *Coxiella burnetii*, a rickettsia, would protect mice against the bacterium *L. monocytogenes* more effectively than would recovery from *L. monocytogenes* itself. Most importantly, our results with haemoprotozoa provided an irrefutable answer to the question of whether these organisms were being killed by enhanced phagocytosis or by the release of soluble mediators from macrophages, with broad-spectrum effects. With our organisms dying inside circulating erythrocytes instead of within macrophages, we could be in no doubt that the mechanism in our case, and conceivably in others as well, was a soluble factor released from macrophages, a concept beginning to emerge in this field (Sharma & Middlebrook, 1977).

TUMOUR NECROSIS FACTOR

Where to head next? The most widespread characteristic of our protectants was that most of them, including BCG (Old, Clarke & Benacerraf, 1959), *C. parvum* (Halpern *et al.* 1966), *Listeria* (Youdin, Moser & Stutman, 1974), *Salmonella* (Hardy & Kotlarski, 1971) and *Coxiella burnetii* (Kelly *et al.* 1976), had been shown to be effective against tumours as well as, in our hands, against protozoa inside red cells. Since tumour cells, on size alone, had almost as good a case as protozoa inside circulating red cells to be resistant to phagocytosis, we became increasingly interested in a mediator termed tumour necrosis factor, or TNF. This had recently been described by Carswell and co-workers at the Sloan Kettering Institute in New York, who were seeking an explanation for the protective effect of BCG and *C. parvum* against experimental tumours (Carswell *et al.* 1975; Helson *et al.* 1975; Green *et al.* 1977). At that time bacterial lipopolysaccharide (LPS) was the only known trigger for TNF release and it also released endogenous pyrogen (subsequently termed interleukin-1). Logically this was the cause of the recurrent fever in malaria, known since last century to begin about 2 hours after the post-schizogony disruption of the red cell that begins each cycle of multiplication of the erythrocytic stage of malaria parasites.

There had been an assumption for about 100 years that schizogony releases some undefined toxin of parasite origin that acted directly on the host to cause illness (reviewed by Kitchen, 1949). We added to this the proposal that such a toxin acted indirectly by inducing host cells to release various harmful mediators, including TNF (Clark, 1978). These were then termed lymphokines and monokines, but as their cellular sources became known to be very diverse they were soon collectively termed cytokines and were realised to be central to the inflammatory response. Since injecting LPS into early-stage malarial mice rapidly (hours) mimicked the pathological changes that normally were not observed until the infections reached their terminal stage, we proposed that these lymphokines and monokines, when produced excessively, caused the fever, hypoglycaemia, bone marrow depression, consumption coagulopathy, hypergammaglobulinemia, hypotension and rise in serum levels of acute phase reactants seen in both endotoxicity and malaria (Clark *et al.* 1981; Clark, 1982*b*). Meanwhile, we had done collaborative experiments with the Sloan Kettering group in which malaria infection proved to prime mice for TNF production as effectively as did BCG or *C. parvum*, and serum containing TNF inhibited *in vivo* multiplication of *Plasmodium vinckei* (Clark *et al.* 1981). Others subsequently reported that rabbit serum rich in TNF inhibited *in vitro* growth of *P. falciparum* (Haidaris *et al.* 1983). We could not detect TNF activity in terminal *P. vinckei* serum, but proposed that this failure was simply because the assays of the day were too insensitive, a prediction subsequently borne out when the advantage of adding actinomycin D to the bioassay became known (Grau *et al.* 1987; Clark & Chaudhri, 1988*b*).

MALARIA AND TNF

In due course recombinant TNF became available and we were able to show that it would reproduce the predicted malaria-like pathology in mice. Much less TNF was required in mice carrying a non-symptomatic low load of malaria parasites (Clark *et al.* 1987*a*), in hindsight because their IFN-γ levels were increased. Likewise, TNF proved to be an excellent emulator of the signs and symptoms of human malaria, in the form of the side-effects caused when it was given therapeutically to tumour patients (Creagan *et al.* 1988; Spriggs *et al.* 1988). By this time TNF had also begun to be recognised as a mediator of the pathology of gram-negative bacterial infections, as we had proposed earlier (Clark *et al.* 1981), and the first experiments with it in this context were being reported (Tracey *et al.* 1986, 1987*a, b*). We were also able to show that recombinant TNF inhibited *in vivo* growth of *P. chabaudi* (Clark *et al.* 1987*b*), and could cause foetal loss (Clark & Chaudhri, 1988*a*) as well as erythrophagocytosis and dyserythropoiesis (Clark & Chaudhri 1988*b*), all of which are associated with malaria infection. Moreover, TNF generation was noted in human monocytes co-cultured with rupturing schizonts (Kwiatkowski *et al.* 1989) and it

began to be found in the circulation of malarial patients in proportion to their severity of illness (Grau *et al.* 1989; Kern *et al.* 1989; Butcher *et al.* 1990; Kwiatkowski *et al.* 1990). Thus, as has been reviewed (Clark, 1987*a, b*; Clark, Chaudhri & Cowden, 1989), TNF escaped from the confines of the tumour world and began to be seen as a *bona fide* mediator of both cell-mediated immunity against malaria parasites and the pathophysiology of the disease itself. Over this period it was realised that TNF was simply one of a number of pro-inflammatory cytokines interacting in these circumstances and was countered by a series of anti-inflammatory cytokines, as well as by soluble forms of the cytokine receptors. New members of the TNF family, such as fas ligand (Helmby, Jonsson & Troye-Blomberg, 2000; Matsumoto *et al.* 2000) have recently appeared on the scene and will no doubt be explored in as much detail as was TNF. All of this adds complexity, but no difference in principle.

MALARIA AND NITRIC OXIDE FROM iNOS

As noted, circulating levels of TNF had been associated with the illness of falciparum malaria, particularly its coma, but this did not immediately suggest how loss of consciousness could occur. Proteins such as these cytokines require a number of subsequent signalling steps before they can influence function and nitric oxide generated by iNOS has come to be recognised as a major candidate for the next step along this pathway. Undoubtedly all of us in this area were greatly influenced by John Hibbs' seminal work on nitric oxide as an effector molecule of macrophage origin (Hibbs *et al.* 1988). The connection of nitric oxide with malarial disease came from linking two observations in unrelated research areas, one that TNF induced nitric oxide release from mammalian endothelial cells (Kilbourn *et al.* 1990) and the other that normal excitatory synaptic activity depended on nitric oxide (Garthwaite, Charles & Chess-Williams, 1988). Since nitric oxide is a non-polar gas that can, like oxygen and carbon dioxide, diffuse freely across cell membranes, it seemed plausible to us that circulating TNF, particularly if concentrated at the site of rupture of sequestered schizonts, could thus influence synaptic function within the brain. Accordingly, we proposed a link, through iNOS-induced nitric oxide, for how TNF could reversibly alter the function of the central nervous system during falciparum malaria (Clark, Rockett & Cowden, 1991; Clark *et al.* 1992*a*).

It has not been easy to establish this connection between nitric oxide and systemic disease, chiefly because of difficulty in assaying for this short-lived molecule, which is active only a very short distance from its cell of origin. Nevertheless, there has been much recent activity in this area, with various groups expressing their interest in the effects of nitric oxide

Table 1. Involvement in killing the infectious agent

Organism	Inflammatory cytokines	Nitric oxide
Plasmodium spp.	Clark *et al.* 1981	Rockett *et al.* 1991
Babesia spp.	Clark, 1979*b*	Rosenblattbin *et al.* 1996
Mycobacterium spp.	Bermudez & Young, 1988	Denis, 1991
Brucella abortus	Zhan *et al.* 1996	Gross *et al.* 1998
Coxiella burnetii	Tokarevich *et al.* 1992	Dellacasagrande *et al.* 1999
Salmonella spp.	Degre & Bukholm, 1990	Meli *et al.* 1996
Listeria monocytogenes	Rothe *et al.* 1993	Bermudez, 1993
Leishmania spp.	Titus *et al.* 1989	Liew *et al.* 1990
Toxoplasma gondii	Chang *et al.* 1990	Chao *et al.* 1993

Table 2. Involvement in causing the disease

Organism	Inflammatory cytokines	Nitric oxide
Plasmodium spp.	Clark *et al.* 1981	Clark *et al.* 1991
Babesia spp.	Clark, 1982*a*	Gale *et al.* 1998
Mycobacterium spp.	Rook *et al.* 1987	Bloom *et al.* 1999
Brucella abortus	Ahmed *et al.* 1999	?
Coxiella burnetii	Mege *et al.* 1997	?
Salmonella spp.	Bhutta *et al.* 1997	MacFarlane *et al.* 1999
Listeria monocytogenes	Nakane *et al.* 1999	MacFarlane *et al.* 1998
Leishmania spp.	Raziuddin *et al.* 1994	Giorgio *et al.* 1996
Toxoplasma gondii	Arsenijevic *et al.* 1997	Khan *et al.* 1997

against both *P. falciparum* (Rockett *et al.* 1991) and the human host, by assaying plasma, urine or cerebrospinal fluid from malaria patients for nitrites and nitrates, the stable oxidation products of nitric oxide (Cot *et al.* 1994; Prada & Kremsner, 1995; Al Yaman *et al.* 1996; Anstey *et al.* 1996; Dondorp *et al.* 1998). Difficulties in the interpretation of these data include estimating how much nitric oxide is actually converted to these anions and knowing precisely where the nitric oxide is being generated. To overcome these obstacles others are employing immunohistochemistry to detect iNOS and nitro-tyrosine, a biochemical footprint for nitric oxide production in autopsy samples.

AN OVERVIEW OF HETEROLOGOUS IMMUNITY

As shown in Tables 1 and 2, the set of principles we developed to help us understand why BCG and *C. parvum* protected against babesia and malaria were, in hindsight, a model for heterologous immunity as a whole. As noted, a literature has developed that implicates TNF and related cytokines, as well as nitric oxide, in the protective host response against, and the disease caused by, the range of organisms noted in this review to be involved in heterologous immunity. Table 1 gives the earliest apparent examples of arguments for the involvement of these

mediators in killing the infectious agent, and Table 2 reports their involvement in generating the disease caused by that organism. The two question marks in the nitric oxide columns are largely because the area has not been explored and at least one group has recorded being unable to demonstrate protection against *Coxiella burnetii* being mediated by nitric oxide (Dellacasagrande *et al.* 1999).

The list of known protective heterologous interactions is still incomplete, with, for example, recent evidence for *Plasmodium vinckei* protecting against subsequent *Salmonella enteritidis* infection (Lehman, Prada & Kremsner, 1998), with nitric oxide being the probable mediator of the effect. This observation has practical implications as well as adding to basic knowledge, since dual infections with these genera are common in the wet tropics. In a recent study in Cameroon, for example, 17% of 200 acute malaria cases were also infected with typhoid (Ammah *et al.* 1999).

IMPLICATIONS FOR VACCINES

Vaccines designed to induce a strong specific antibody-based immune response against the organisms listed in the tables have, at best, not been very effective or long lasting. With some of these parasites it is accepted that this is because the immunity that develops during natural infections is cell mediated, another way of saying that they rely on the same mediators as does heterologous immunity. Observations that led to the argument that children exposed to *Plasmodium vivax* have a degree of protection against *P. falciparum* (Maitland, Williams & Newbold, 1997) may have this mechanism as part of their explanation. The steps along the pathway of antigen recognition to effector molecule are, however, still imperfectly understood, as is its potency compared to humoral immunity. Research to develop and refine vaccines based on this cell-mediated immunity is still hampered by insufficient knowledge of how they actually work, combined with an inability, to date, to develop *in vitro* or *in vivo* correlates of their potency. The usual assumption is that this response is a primitive initial mechanism that protects the host until specific immunity, typically based on antibody, gets underway. But this is often untested and the possibility that the non-specific system is in fact never superseded warrants closer investigation in some of these parasites.

Probably the best known example of how strong heterologous immunity can be is that seen in *B. microti* in the mouse. Certainly no mechanism of immunity stronger or more durable than that induced in this model by BCG, killed *C. parvum* or *C. burnetii* extract would ever be needed for absolute protection against an invading organism. For example the protection BCG gives in this circumstance has been shown to be fully active for many months, to protect against a massive challenge (10^9 organisms) and to do so through a mechanism that causes intra-erythrocytic pyknosis of the challenge dose of parasites in the absence of antibody (Clark *et al.* 1977*b*). It also protects absolutely, with parasites beginning to die inside red cells within hours of injection, against 10^9 *Babesia rodhaini*, which is invariably fatal in normal mice when just one parasite is injected. As these authors noted, all of these findings are characteristic of natural immunity against this parasite. Thus it is plausible that no specific antibody-based immune mechanism ever becomes dominant in *B. microti*-infected mice and that generating a vaccine on these principles would be correspondingly difficult. In this way, the susceptibility of a parasite to heterologous immunity could well be a marker of its susceptibility to cell-mediated compared to humoral immunity, useful information when planning to develop a vaccine against it. For example, BCG protected less solidly against *Plasmodium yoelii* than against *B. microti* (Clark *et al.* 1976) and humoral immune responses appear to play a correspondingly larger role (Matsumoto *et al.* 1998).

By this circuitous route, the heterologous immunity reported between babesiosis and malaria in mice (Cox, 1970) has led to the cytokine theory of disease pathogenesis and has also uncovered the role of these mediators in the host response against a wide range of infectious organisms. It also has provided useful background information for vaccine researchers. This will continue to be a growth area in parasitology.

REFERENCES

AHMED, K., AL MATROUK, K. A., MARTINEZ, G., OISHI, K., ROTIMI, V. O. & NAGATAKE, T. (1999). Increased serum levels of interferon-gamma and interleukin-12 during human brucellosis. *American Journal of Tropical Medicine and Hygiene* **61**, 425–427.

AL YAMAN, F., GENTON, B., MOKELA, D., ROCKETT, K. A., ALPERS, M. P. & CLARK, I. A. (1996). Association between serum levels of reactive nitrogen intermediates and coma in children with cerebral malaria in Papua New Guinea. *Transactions of the Royal Society of Tropical Medicine and Hygiene* **90**, 270–273.

AMMAH, A., NKUO, A. T., NDIP, R. & DEAS, J. E. (1999). An update on concurrent malaria and typhoid fever in Cameroon. *Transactions of the Royal Society of Tropical Medicine and Hygiene* **93**, 127–129.

ANSTEY, N. M., WEINBERG, J. B., HASSANALI, M., MWAIKAMBO, E. D., MANYENGA, D., MISUKONIS, M. A., ARNELLE, D. R., HOLLIS, D., MCDONALD, M. I. & GRANGER, D. L. (1996). Nitric oxide in Tanzanian children with malaria: inverse relationship between malaria severity and nitric oxide production/nitric oxide synthase type 2 expression. *Journal of Experimental Medicine* **184**, 557–567.

ARSENIJEVIC, D., GIRARDIER, L., SEYDOUX, J., CHANG, H. R. & DULLOO, A. G. (1997). Altered energy balance and cytokine gene expression in a murine model of chronic infection with *Toxoplasma gondii*. *American Journal of Physiology* **272**, E908–E917.

BERMUDEZ, L. E. (1993). Differential mechanisms of intracellular killing of *Mycobacterium avium* and *Listeria monocytogenes* by activated human and murine macrophages. The role of nitric oxide. *Clinical and Experimental Immunology* **91**, 277–281.

BERMUDEZ, L. E. & YOUNG, L. S. (1988). Tumor necrosis factor, alone or in combination with IL-2, but not IFN-gamma, is associated with macrophage killing of *Mycobacterium avium* complex. *Journal of Immunology* **140**, 3006–3013.

BHUTTA, Z. A., MANSOORALI, N. & HUSSAIN, R. (1997). Plasma cytokines in paediatric typhoidal salmonellosis: correlation with clinical course and outcome. *Journal of Infection* **35**, 253–256.

BLOOM, B. R., MAZZACCARO, R. J., FLYNN, J. A., CHAN, J., SOUSA, A., SALGAME, P., STENGER, S., MODLIN, R. L., KRENSKY, A., DEMANT, P. & KRAMNI, I. (1999). Immunology of an infectious disease: Pathogenesis and protection in tuberculosis. *Immunologist* **7**, 54–59.

BURGNER, D., XU, W. M., ROCKETT, K., GRAVENOR, M., CHARLES, I. G., HILL, A. V.S. & KWIATKOWSKI, D. (1998). Inducible nitric oxide synthase polymorphism and fatal cerebral malaria. *Lancet* **352**, 1193–1194.

BUTCHER, G. A., GARLAND, T., ADJUKIEWICZ, A. B. & CLARK, I. A. (1990). Serum TNF associated with malaria in patients in the Solomon Islands. *Transactions of the Royal Society of Tropical Medicine and Hygiene* **85**, 658–661.

CARSWELL, E. A., OLD, L. J., KASSEL, R. L., GREEN, S., FIORE, N. & WILLIAMSON, B. (1975). An endotoxin induced serum factor that causes necrosis of tumors. *Proceedings of the National Academy of Sciences, USA* **72**, 3666–3670.

CHANG, H. R., GRAU, G. E. & PERCHERE, J. C. (1990). Role of TNF and IL-1 in infections with *Toxoplasma gondii. Immunology* **69**, 33–37.

CHAO, C. C., ANDERSON, W. R., HU, S., GEKKER, G., MARTELLA, A. & PETERSON, P. K. (1993). Activated microglia inhibit multiplication of *Toxoplasma gondii* via a nitric oxide mechanism. *Clinical Immunology and Immunopathology* **67**, 178–183.

CLARK, I. A. (1978). Does endotoxin cause both the disease and parasite death in acute malaria and babesiosis? *Lancet* **ii**, 75–77.

CLARK, I. A. (1979*a*). Protection of mice against *Babesia microti* with cord factor, COAM, zymosan, glucan, *Salmonella* and *Listeria. Parasite Immunology* **1**, 179–196.

CLARK, I. A. (1979*b*). Resistance to *Babesia* spp. and *Plasmodium* sp. in mice pretreated with an extract of *Coxiella burnetii. Infection and Immunity* **24**, 319–325.

CLARK, I. A. (1982*a*). Correlation between susceptibility to malaria and babesia parasites and to endotoxin. *Transactions of the Royal Society of Tropical Medicine and Hygiene* **76**, 4–7.

CLARK, I. A. (1982*b*). Suggested importance of monokines in pathophysiology of endotoxin shock and malaria. *Klinische Wochenschrift* **60**, 756–758.

CLARK, I. A. (1987*a*). Cell-mediated immunity in protection and pathology of malaria. *Parasitology Today* **3**, 300–305.

CLARK, I. A. (1987*b*). Monokines and lymphokines in malarial pathology. *Annals of Tropical Medicine and Parasitology* **81**, 577–585.

CLARK, I. A., ALLISON, A. C. & COX, F. E. G. (1976). Protection of mice against *Babesia* and *Plasmodium* with BCG. *Nature* **259**, 309–311.

CLARK, I. A. & CHAUDHRI, G. (1988*a*). Tumor necrosis factor in malaria-induced abortion. *American Journal of Tropical Medicine and Hygiene* **39**, 246–249.

CLARK, I. A. & CHAUDHRI, G. (1988*b*). Tumour necrosis factor may contribute to the anaemia of malaria by causing dyserythropoiesis and erythrophagocytosis. *British Journal of Haematology* **70**, 99–103.

CLARK, I. A., CHAUDHRI, G. & COWDEN, W. B. (1989). Roles of tumour necrosis factor in the illness and pathology of malaria. *Transactions of the Royal Society of Tropical Medicine and Hygiene* **83**, 436–440.

CLARK, I. A., COWDEN, W. B., BUTCHER, G. A. & HUNT, N. H. (1987*a*). Possible roles of tumor necrosis factor in the pathology of malaria. *American Journal of Pathology* **129**, 192–199.

CLARK, I. A., COX, F. E. G. & ALLISON, A. C. (1977*a*). Protection of mice against *Babesia* spp. and *Plasmodium* spp. with killed *Corynebacterium parvum. Parasitology* **74**, 9–18.

CLARK, I. A., HUNT, N. H., BUTCHER, G. A. & COWDEN, W. B. (1987*b*). Inhibition of murine malaria (*Plasmodium chabaudi*) *in vivo* by recombinant interferon-gamma or tumor necrosis factor, and its enhancement by butylated hydroxyanisole. *Journal of Immunology* **139**, 3493–3496.

CLARK, I. A., MacMICKING, J. D., GRAY, K. M., ROCKETT, K. A. & COWDEN, W. B. (1992*a*). Malaria mimicry with tumor necrosis factor – contrasts between species of murine malaria and *Plasmodium falciparum. American Journal of Pathology* **140**, 325–336.

CLARK, I. A., RICHMOND, J. E., WILLS, E. J. & ALLISON, A. C. (1975). Immunity to intra-erythrocytic protozoa. *Lancet* **ii**, 1128–1129.

CLARK, I. A., ROCKETT, K. A. & COWDEN, W. B. (1991). Proposed link between cytokines, nitric oxide, and human cerebral malaria. *Parasitology Today* **7**, 205–207.

CLARK, I. A., ROCKETT, K. A. & COWDEN, W. B. (1992*b*). Possible central role of nitric oxide in conditions clinically similar to cerebral malaria. *Lancet* **340**, 894–896.

CLARK, I. A., VIRELIZIER, J.-L., CARSWELL, E. A. & WOOD, P. R. (1981). Possible importance of macrophage-derived mediators in acute malaria. *Infection and Immunity* **32**, 1058–1066.

CLARK, I. A., WILLS, E. J., RICHMOND, J. E. & ALLISON, A. C. (1977*b*). Suppression of babesiosis in BCG-infected mice and its correlation with tumor inhibition. *Infection and Immunity* **17**, 430–438.

COLLINS, F. M. & SCOTT, M. T. (1974). Effect of *Corynebacterium parvum* on the growth of *Salmonella enteritidis* in mice. *Infection and Immunity* **9**, 863–869.

COT, S., RINGWALD, P., MULDER, B., MIAILHES, P., YAPYAP, J., NUSSLER, A. K. & ELING, W. M. C. (1994). Nitric oxide in cerebral malaria. *Journal of Infectious Diseases* **169**, 1417–1418.

COX, F. E. G. (1970). Protective immunity between malaria parasites and piroplasms in mice. *Bulletin of the World Health Organisation* **43**, 325–336.

COX, F. E. G. (1978). Heterologous immunity between piroplasms and malaria parasites: the simultaneous elimination of *Plasmodium vinckei* and *Babesia microti* from the blood of doubly infected mice. *Parasitology* **76**, 55–60.

CREAGAN, E. T., KOVACH, J. S., MOERTEL, C. G., FRYTAK, S. & KVOLS, L. K. (1988). A phase 1 clinical trial of recombinant human tumor necrosis factor. *Cancer* **62**, 2467–2471.

DEGRE, M. & BUKHOLM, G. (1990). Effect of tumor necrosis factor-alpha on infection with *Salmonella typhimurium* in a mouse model. *Journal of Biological Regulators and Homeostatic Agents* **4**, 157–161.

DELLACASAGRANDE, J., CAPO, C., RAOULT, D. & MEGE, J. L. (1999). IFN-gamma-mediated control of *Coxiella burnetii* survival in monocytes: The role of cell apoptosis and TNF. *Journal of Immunology* **162**, 2259–2265.

DENIS, M. (1991). Interferon-gamma-treated murine macrophages inhibit growth of tubercle bacilli via the generation of reactive nitrogen intermediates. *Cellular Immunology* **132**, 150–157.

DONDORP, A. M., PLANCHE, T., DEBEL, E. E., ANGUS, B. J., CHOTIVANICH, K. T., SILAMUT, K., ROMIJN, J. A., RUANGVEERAYUTH, R., HOEK, F. J., KAGER, P. A., VREEKEN, J. & WHITE, N. J. (1998). Nitric oxides in plasma, urine, and cerebrospinal fluid in patients with severe falciparum malaria. *American Journal of Tropical Medicine and Hygiene* **59**, 497–502.

DUBOS, R. J. & SCHAEDLER, R. W. (1957). Effect of cellular constituents of *Mycobacteria* on the resistance of mice to heterologous infection. *Journal of Experimental Medicine* **106**, 703–709.

GALE, K. R., WALTISBUHL, D. J., BOWDEN, J. M., JORGENSEN, W. K., MATHESON, J., EAST, I. J., ZAKRZEWSKI, H. & LEATCH, G. (1998). Amelioration of virulent *Babesia bovis* infection in calves by administration of the nitric oxide synthase inhibitor aminoguanidine. *Parasite Immunology* **20**, 441–445.

GARTHWAITE, J., CHARLES, S. L. & CHESS-WILLIAMS, R. (1988). Endothelium-derived relaxing factor release on activation of the NMDA receptors suggests role as intercellular messenger in the brain. *Nature* **336**, 385–388.

GIORGIO, S., LINARES, E., CAPURRO, M. D. L., DE BIANCHI, A. G. & AUGUSTO, O. (1996). Formation of nitrosyl haemoglobin and nitrotyrosine during murine leishmaniasis. *Photochemistry and Photobiology* **63**, 750–754.

GRAU, G. E., FAJARDO, L. F., PIQUET, P.-F., ALLET, B., LAMBERT, P.-H. & VASSALI, P. (1987). Tumor necrosis factor (cachectin) as an essential mediator in murine cerebral malaria. *Science* **237**, 1210–1212.

GRAU, G. E., TAYLOR, T. E., MOLYNEUX, M. E., WIRIMA, J. J., VASSALLI, P., HOMMEL, M. & LAMBERT, P.-H. (1989). Tumor necrosis factor and disease severity in children with falciparum malaria. *New England Journal of Medicine* **320**, 1586–1591.

GREEN, S., DOBRJANSKY, A., CHIASSON, M. A., CARSWELL, E., SCHWARTZ, M. K. & OLD, L. J. (1977). *Corynebacterium parvum* as the priming agent in the production of tumor necrosis factor in the mouse. *Journal of the National Cancer Institute* **59**, 1519–1522.

GROSS, A., SPIESSER, S., TERRAZA, A., ROUOT, B., CARON, E. & DORNAND, J. (1998). Expression and bactericidal activity of nitric oxide synthase in *Brucella suis*-infected murine macrophages. *Infection and Immunity* **66**, 1309–1316.

HAIDARIS, C. G., HAYNES, J. D., MELTZER, M. S. & ALLISON, A. C. (1983). Serum containing tumor necrosis factor is cytotoxic for the human malarial parasite *Plasmodium falciparum*. *Infection and Immunity* **42**, 385–393.

HALPERN, B. N., BIOZZI, G., STIFFEL, C. & MOUTON, D. (1966). Inhibition of tumour growth by administration of killed *Corynebacterium parvum*. *Nature* **212**, 853–854.

HARDY, D. & KOTLARSKI, I. (1971). Resistance of mice to Erhlich ascites tumour after immunisation of mice with live *Salmonella interitidis* 11RX. *Australian Journal of Experimental Biology and Medical Science* **49**, 271–279.

HELMBY, H., JONSSON, G. & TROYE-BLOMBERG, M. (2000). Cellular changes and apoptosis in the spleens and peripheral blood of mice infected with blood-stage *Plasmodium chabaudi chabaudi* AS. *Infection and Immunity* **68**, 1485–1490.

HELSON, L., GREEN, S., CARSWELL, E. & OLD, L. J. (1975). Effect of tumour necrosis factor on cultured human melanoma cells. *Nature* **258**, 731–732.

HEROD, E., CLARK, I. A. & ALLISON, A. C. (1978). Protection of mice against the haemoprotozoan *Babesia microti* with *Brucella abortus* strain 19. *Clinical and Experimental Immunology* **31**, 518–523.

HIBBS, J. B., TAINTOR, R. R., VAVRIN, Z. & RACHLIN, E. M. (1988). Nitric oxide: a cytotoxic activated macrophage effector molecule. *Biochemical and Biophysical Research Communications* **157**, 87–94.

HOWARD, J. G., BIOZZI, G., HALPERN, B. N., STIFFEL, C. & MOUTON, D. (1959). The effect of *Mycobacterium tuberculosis* (BCG) infection on the resistance of mice to bacterial endotoxin and *Salmonella enteritidis* infection. *British Journal of Experimental Pathology* **40**, 281–290.

JESPERSEN, A. (1976). Acquired resistance of BCG-vaccinated red mice to infection with *Listeria monocytogenes*. *Acta Pathologica Microbiologica Scandinavica* **84**, 265–272.

KELLY, M. T., GRANGER, D. L., RIBI, E., MILNER, K. C., STRAIN, S. M. & STOENNER, H. G. (1976). Tumor regression with Q fever rickettsia and a mycobacterial glycolipid. *Cancer Immunology and Immunotherapy* **1**, 187–191.

KERN, P., HEMMER, C. J., VAN DAMME, J., GRUSS, H.-J. & DIETRICH, M. (1989). Elevated tumour necrosis factor alpha and interleukin-6 serum levels as markers for complicated *Plasmodium falciparum* malaria. *American Journal of Medicine* **87**, 139–143.

KHAN, I. A., SCHWARTZMAN, J. D., MATSUURA, T. & KASPER, L. H. (1997). A dichotomous role for nitric oxide during acute *Toxoplasma gondii* infection in mice. *Proceedings of the National Academy of Sciences, USA* **94**, 13955–13960.

KILBOURN, R. G., GROSS, S., JUBRAN, A., GRIFFITH, O. W., LEVI, R. & LODATO, R. F. (1990). N^G-methyl-L-arginine inhibits tumor necrosis factor-induced hypotension: implications for the involvement of nitric oxide. *Proceedings of the National Academy of Sciences, USA* **87**, 3629–3632.

KITCHEN, S. F. (1949). Falciparum malaria. In *Malariology* (ed. Boyd, M. F.), pp. 966–994. Philadelphia, W. B. Saunders.

KNIGHT, J. C., UDALOVA, I., HILL, A. V. S., GREENWOOD, B. M., PESHU, N., MARSH, K. & KWIATKOWSKI, D. (1999). A polymorphism that affects OCT-1 binding to the TNF promoter region is associated with severe malaria. *Nature Genetics* **22**, 145–150.

KUN, J. F. J., MORDMULLER, B., LELL, B., LEHMAN, L. G., LUCKNER, D. & KREMSNER, P. G. (1998). Polymorphism in promoter region of inducible nitric oxide synthase gene and protection against malaria. *Lancet* **351**, 265–266.

KWIATKOWSKI, D., CANON, J. G., MANOGUE, K. R., CERAMI, A., DINARELLO, C. A. & GREENWOOD, B. M. (1989). Tumor necrosis factor production in falciparum malaria and its association with schizont rupture. *Clinical and Experimental Immunology* **77**, 361–366.

KWIATKOWSKI, D., HILL, A. V. S., SAMBOU, I., TWUMASI, P., CASTRACANE, J., MANOGUE, K. R., CERAMI, A., BREWSTER, D. R. & GREENWOOD, B. M. (1990). TNF concentration in fatal cerebral, non-fatal cerebral, and uncomplicated *Plasmodium falciparum* malaria. *Lancet* **336**, 1201–1204.

LEHMAN, L. G., PRADA, J. & KREMSNER, P. G. (1998). Protection of mice previously infested with *Plasmodium vinckei* against subsequent *Salmonella enteritidis* infection is associated with nitric oxide production capacity. *Parasitology Research* **84**, 63–68.

LIEW, F. Y., MILLOTT, S., PARKINSON, C., PALMER, R. M. & MONCADA, S. (1990). Macrophage killing of *Leishmania* parasite *in vivo* is mediated by nitric oxide from L-arginine. *Journal of Immunology* **144**, 4794–4797.

MacFARLANE, A. S., HUANG, D., SCHWACHA, M. G., MEISSLER, J. J., GAUGHAN, J. P. & EISENSTEIN, T. K. (1998). Nitric oxide mediates immunosuppression induced by *Listeria monocytogenes* infection: quantitative studies. *Microbial Pathogenesis* **25**, 267–277.

MacFARLANE, A. S., SCHWACHA, M. G. & EISENSTEIN, T. K. (1999). *In vivo* blockage of nitric oxide with aminoguanidine inhibits immunosuppression induced by an attenuated strain of *Salmonella typhimurium*, potentiates *Salmonella* infection, and inhibits macrophage and polymorphonuclear leukocyte influx into the spleen. *Infection and Immunity* **67**, 891–898.

MAEGRAITH, B. (1948). *Pathological Process in Malaria and Blackwater Fever*. Oxford, Blackwell.

MAITLAND, K., WILLIAMS, T. N. & NEWBOLD, C. I. (1997). *Plasmodium vivax* and *P. falciparum*: biological interactions and the possibility of cross-species immunity. *Parasitology Today* **13**, 227–231.

MARSH, K., ENGLISH, M., CRAWLEY, J. & PESHU, N. (1996). The pathogenesis of severe malaria in African children. *Annals of Tropical Medicine and Parasitology* **90**, 395–402.

MATSUMOTO, J., KAWAI, S., TERAO, K., KIRINOKI, M., YASUTOMI, Y., AIKAWA, M. & MATSUDA, H. (2000). Malaria infection induces rapid elevation of the soluble fas ligand level in serum and subsequent T lymphocytopenia: Possible factors responsible for the differences in susceptibility of two species of Macaca monkeys to *Plasmodium coatneyi* infection. *Infection and Immunity* **68**, 1183–1188.

MATSUMOTO, S., YUKITAKE, H., KANBARA, H. & YAMADA, T. (1998). Recombinant *Mycobacterium bovis* Bacillus Calmette-Guérin secreting merozoite surface protein 1 (MSP1) induces protection against rodent malaria parasite infection depending on MSP1-stimulated interferon-gamma and parasite-specific antibodies. *Journal of Experimental Medicine* **188**, 845–854.

McGUIRE, W., KNIGHT, J. C., HILL, A. V. S., ALLSOPP, C. E. M., GREENWOOD, B. M. & KWIATKOWSKI, D. (1999). Severe malarial anemia and cerebral malaria are associated with different tumor necrosis factor promoter alleles. *Journal of Infectious Diseases* **179**, 287–290.

MEGE, J. L., MAURIN, M., CAPO, C. & RAOULT, D. (1997). *Coxiella burnetii*: the 'query' fever bacterium – a model of immune subversion by a strictly intracellular microorganism. *FEMS Microbiology Reviews* **19**, 209–217.

MELI, R., RASO, G. M., BENTIVOGLIO, C., NUZZO, I., GALDIERO, M. & DI CARLO, R. (1996). Recombinant human prolactin induces protection against *Salmonella typhimurium* infection in the mouse: role of nitric oxide. *Immunopharmacology* **34**, 1–7.

NAKANE, A., YAMADA, K., HASEGAWA, S., MIZUKI, D., MIZUKI, M., SASAKI, S. & MIURA, T. (1999). Endogenous cytokines during a lethal infection with *Listeria monocytogenes* in mice. *FEMS Microbiology Letters* **175**, 133–142.

NUMATA, M., SUZUKI, S., MIYAZAWA, N., MIYASHITA, A., NAGASHIMA, Y., INOUE, S., KANEKO, T. & OKUBO, T. (1998). Inhibition of inducible nitric oxide synthase prevents LPS-induced acute lung injury in dogs. *Journal of Immunology* **160**, 3031–3037.

NYKA, W. (1957). Enhancement of resistance to turberculosis in mice experimentally infected with *B. abortus*. *American Review of Tuberculosis* **73**, 251–257.

OLD, L. J., CLARKE, D. A. & BENACERRAF, B. (1959). Effect of Bacillus Calmette Guérin infection on transplanted tumours in the mouse. *Nature* **184**, 291–292.

PRADA, J. & KREMSNER, P. G. (1995). Enhanced production of reactive nitrogen intermediates in human and murine malaria. *Parasitology Today* **11**, 409–410.

RAZIUDDIN, S., ABDALLA, R. E., EL AWAD, E. H. & AL JANADI, M. (1994). Immunoregulatory and proinflammatory cytokine production in visceral and cutaneous leishmaniasis. *Journal of Infectious Diseases* **170**, 1037–1040.

ROCKETT, K. A., AWBURN, M. M., COWDEN, W. B. & CLARK, I. A. (1991). Killing of *Plasmodium falciparum in vitro* by nitric oxide derivatives. *Infection and Immunity* **59**, 3280–3283.

ROOK, G. A. W., TAVERNE, J., LEVETON, C. & STEELE, J. (1987). The role of gamma-interferon, vitamin D_3 metabolites and tumour necrosis factor in the pathogenesis of tuberculosis. *Immunology* **62**, 229–234.

ROSENBLATTBIN, H., KLEIN, A. & SREDNI, B. (1996). Antibabesial effect of the immunomodulator AS101 in mice: role of increased production of nitric oxide. *Parasite Immunology* **18**, 297–306.

ROTHE, J., LESSLAUER, W., LOTSCHER, H., LANG, Y., KOEBEL, P., KONTGEN, F., ALTHAGE, A., ZINKERNAGEL, R., STEINMETZ, M. & BLUETHMANN, H. (1993). Mice lacking the tumour necrosis factor receptor-1 are resistant to TNF-Mediated toxicity but highly susceptible to infection by *Listeria monocytogenes*. *Nature* **364**, 798–802.

RUETTEN, H., SOUTHAN, G. J., ABATE, A. & THIEMERMANN, C. (1996). Attenuation of endotoxin-induced multiple organ dysfunction by 1-amino-2-hydroxy-guanidine, a potent inhibitor of inducible nitric oxide synthase. *British Journal of Pharmacology* **118**, 261–270.

SALVIN, S. B., RIBI, E., GRANGER, D. L. & YOUNGNER, J. S. (1975). Migration inhibitory factor and type II interferon in the circulation of mice sensitized with mycobacterial components. *Journal of Immunology* **114**, 354–359.

SCHWARTZ, D., MENDONCA, M., SCHWARTZ, I., XIA, Y. Y., SATRIANO, J., WILSON, C. B. & BLANTZ, R. C. (1997). Inhibition of constitutive nitric oxide synthase (NOS) by nitric oxide generated by inducible NOS after lipopolysaccharide administration provokes renal dysfunction in rats. *Journal of Clinical Investigation* **100**, 439–448.

SHARMA, S. D. & MIDDLEBROOK, G. (1977). Antibacterial product of peritoneal exudate cell cultures from guinea pigs infected with Mycobacteria, Listeriae, and Rickettsiae. *Infection and Immunity* **15**, 745–750.

SMRKOVSKI, L. L. & LARSON, C. L. (1977). Effect of treatment with BCG on the course of visceral leishmaniasis in BALB/c mice. *Infection and Immunity* **16**, 249–257.

SPRIGGS, D. R., SHERMAN, M. L., MICHIE, H., ARTHUR, K. A., IMAMURA, K., WILMORE, D., FREI, E. & KUFE, D. W. (1988). Recombinant human tumor necrosis factor administered as a 24-hour intravenous infusion. A phase 1 and pharmacologic study. *Journal of the National Cancer Institute* **80**, 1039–1044.

SWARTZBERG, J. E., KRAHENBUHL, J. L. & REMINGTON, J. S. (1975). Dichotomy between macrophage activation and degree of protection against *Listeria monocytogenes* and *Toxoplasma gondii* in mice stimulated with *Corynebacterium parvum*. *Infection and Immunity* **12**, 1037–1043.

TALIAFERRO, W. H. & CANNON, P. R. (1936). The cellular reactions during primary infections and super-infections of *Plasmodium brasilianum* in Panamanian monkeys. *Journal of Infectious Diseases* **59**, 72–83.

TAYLOR, T. E., BORGSTEIN, A. & MOLYNEUX, M. E. (1993). Acid-base status in paediatric *Plasmodium falciparum* malaria. *Quarterly Journal of Medicine* **86**, 99–109.

TITUS, R. G., SHERRY, B. & CERAMI, A. (1989). Tumor necrosis factor plays a protective role in experimental murine cutaneous leishmaniasis. *Journal of Experimental Medicine* **170**, 2097–2104.

TOKAREVICH, N. K., PROKOPYEV, A. A., PROKOPYEVA, E. D., SIMBIRTSEV, A. S., TOROPOVA, B. G., DAITER, A. B. & KETLINSKY, S. A. (1992). Role of tumor necrosis factor and interleukin-1 in the formation of resistance in experimental Q fever. *Zhurnal Mikrobiologii Epidemiologii I Immunobiologii* **5**, 46–47.

TRACEY, K. J., BEUTLER, B., LOWRY, S. F., MERRYWEATHER, J., WOLPE, S., MILSARK, I. W., HARIRI, R. J., FAHEY, T. J., ZENTELLA, A., ALBERT, J. D., SHIRES, G. T. & CERAMI, A. (1986). Shock and tissue injury induced by recombinant human cachectin. *Science* **234**, 470–474.

TRACEY, K. J., FONG, Y., HESSE, D. G., MANOGUE, K. R., LEE, A. T., KUO, G. C., LOWRY, S. F. & CERAMI, A. (1987 *a*). Anti-cachectin/TNF monoclonal antibodies prevent septic shock during lethal shock bacteraemia. *Nature* **330**, 662–664.

TRACEY, K. J., LOWRY, S. F., FAHEY, T. J., ALBERT, J. D., FONG, Y., HESSE, D., BEUTLER, B., MANOGUE, K. R., CALVANO, S., CERAMI, A. & SHIRES, G. T. (1987 *b*). Cachectin/tumor necrosis factor induces lethal shock and stress hormone response in the dog. *Surgery, Gynecology and Obstetrics* **164**, 415–422.

YOUDIN, S., MOSER, M. & STUTMAN, O. (1974). Non-specific suppression of tumour growth by an immune reaction to *Listeria monocytogenes*. *Journal of the National Cancer Institute* **52**, 193–198.

ZHAN, Y. F., LIU, Z. Q. & CHEERS, C. (1996). Tumor necrosis factor alpha and interleukin-12 contribute to resistance to the intracellular bacterium *Brucella abortus* by different mechanisms. *Infection and Immunity* **64**, 2782–2786.

ZINKERNAGEL, R. M. (1976). Cell-mediated immune response to *Salmonella typhimurium* infection in mice: development of nonspecific bacteriocidal activity against *Listeria monocytogenes*. *Infection and Immunity* **13**, 1069–1073.

Use of an optimality model to solve the immunological puzzle of concomitant infection

A. L. GRAHAM*

318 Corson Hall, Department of Ecology and Evolutionary Biology, Cornell University, Ithaca, New York 14853, U.S.A.

SUMMARY

Immunological data indicate that different subsets of T-helper cells work best against different types of infection. Concomitant infection of a host may thus impose either conflicting or synergistic immune response requirements, depending upon the extent to which the component optimal immune responses differ. Drawing upon empirically-determined optimal responses to single-species infections, an optimality model is here used to generate testable hypotheses for optimal responses to concomitant infection. The model is based upon the principle that the joint immune response will minimize divergence from each of the optima for single-species infections, but that it will also be weighted by the importance of mounting the correct response against each infectious organism. The model thus predicts a weighted average response as the optimal response to concomitant infection. Data on concomitant infection of murine hosts by the parasites *Schistosoma mansoni* and *Toxoplasma gondii* will provide the first test of the optimality model. If the weighted average hypothesis holds true, then there are no emergent immunological properties of concomitant infections and we may be able to understand immune responses to concomitant infection directly via our understanding of single-species infections.

Key words: cytokine, T-helper cell, optimal response, concomitant infection, Schistosoma mansoni, Toxoplasma gondii
.

INTRODUCTION

Concomitant infection of a host by disparate parasite or pathogen species can pose a multi-pronged challenge for the immune system: how should the cells and molecules of the immune system be allocated amongst the simultaneous responses required by concomitant infection? Experimental immunology has provided a wealth of data on the optimal T-helper cell responses to single-species infections (reviewed in Mosmann & Sad, 1996; Romagnani, 1996; Infante-Duarte & Kamradt, 1999), and, to some extent, on responses to dual infections (e.g. Marshall *et al.* 1999). Some theoretical work has explored the proliferation of T-helper cells in response to infection (e.g. Morel, Kalagnanam & Morel, 1992; Schweitzer, Swinton & Anderson, 1993; Fishman & Perelson, 1999). There has, however, been no attempt to synthesize the empirical data into a general theory of the optimal allocation of T-helper cells among concomitant infections. The optimality model presented here represents just such an attempt at synthesis. Specifically, the question that the model poses is: can the best T-helper response to concomitant infection (i.e. the response that minimizes host morbidity) be predicted from the empirically-determined optimal T-helper responses to each of the component infections? A test of the model against empirical data is under way.

* Tel: (607) 254-4296. Fax: (607) 255-8088. E-mail: alg13@cornell.edu

The induced subset of T-helper cells determines which effector mechanisms will act against infection, so these cells (and their associated cytokines) have an important role in determining immune response efficacy (Infante-Duarte & Kamradt, 1999; Janeway *et al.* 1999). Generally speaking, for each parasite or pathogen, there is a corresponding 'optimal' T-helper/cytokine response that fights infection most efficiently and with least pathology. T-helper type 1 (Th_1) cells produce pro-inflammatory cytokines and mediate cytotoxicity by effectors such as killer T cells and macrophages. T-helper type 2 (Th_2) cells, on the other hand, produce cytokines that help to stimulate antibody production by B cells and to recruit effectors such as eosinophils. Th_1 cells generally work best against intracellular parasites/pathogens, whereas Th_2 cells tend to generate effective responses against extracellular parasites/pathogens (Sher & Coffman, 1992; Mosmann & Sad, 1996; Romagnani, 1996) (but see Allen & Maizels, 1997). The empirically-determined optimal T-helper response to a single-species infection is here defined as the response that maximizes host survival or minimizes morbidity in mouse model systems.

The T-helper response might likewise determine the outcome of concomitant infection. Importantly, the T-helper subsets tend to be mutually down-regulatory (Sher & Coffman, 1992; Janeway *et al.* 1999); that is, a Th_1 cell produces cytokines that discourage its neighbours from becoming Th_2 cells, and vice versa. It is therefore difficult for the immune system to 'do both', at least within a given site, and conflicts among T-helper response requirements

Parasitology (2001), **122**, S61–S64.   Printed in the United Kingdom   © 2001 Cambridge University Press

might arise when different types of responses are required simultaneously. Empirical data suggest that concomitant infection can indeed lead to T-helper response synergies or conflicts, depending on which effector mechanisms are required to clear each infection (Brunet, Dunne & Pearce, 1998). Response synergies (e.g. the induction of Th_2-dominated responses by both schistosomes and intestinal nematodes) may decrease the pathological impact of concomitant infection (Curry *et al.* 1995). Response conflicts (e.g. simultaneous requirements for a Th_1-dominated response against *Toxoplasma* and a Th_2-dominated response against *Schistosoma*), on the other hand, might exacerbate disease (Marshall *et al.* 1999). Given this potential for T-helper synergy or conflict, how are T-helper cells allocated in a host infected with multiple parasite/pathogen species? Are they allocated in a manner that may be predicted from the optimal responses to each component (single-species) infection?

An optimality model is a useful analytical tool for the exploration of resource allocation problems in biology (Parker & Maynard Smith, 1990). The model proposed here, grounded in empirical data from experimental immunology, generates testable hypotheses about optimal T-helper responses to concomitant infection. Given 'local' (single-infection) immune response optima from the empirical database, the model proposes a 'global' optimization scheme for the case of concomitant infection. Whether or not this simple model can accurately predict the best T-helper response to concomitant infection, the process of testing optimality hypotheses will inform our understanding of the complexities of the immune system.

MODELLING METHODS AND RESULTS

Summarizing the immune response in one variable

The first step in the modelling process was to formulate a quantitative summary of the complex network of interactions that comprise an immune response. Summarizing an immune response in one variable allows quantitative comparisons among immune responses, across hosts and across infections. Because immune responses actually fall out along a continuum of cytokine and effector combinations rather than into a qualitative (Th_1 versus Th_2) dichotomy (Allen & Maizels, 1997), a continuous variable may indeed be the best way to characterise a T-helper response to infection.

Here, the summary variable t is defined as the proportional representation of T-helper type 1-associated cytokines in the total cytokine pool. Focusing upon interferon-gamma (IFN-γ) as a driver of Th_1 responses and interleukin 4 (IL-4) as a driver of Th_2 responses, the systemic T-helper response may be quantified in terms of their relative concentrations:

$$t = \frac{[\text{IFN}-\gamma]}{[\text{IFN}-\gamma]+[\text{IL}-4]}. \tag{1}$$

An IFN-γ-driven (Th_1) response will make t approach 1, whereas an IL-4-driven (Th_2) response will make t approach 0. The variable t is thus a measure of the extent to which the immune response is skewed toward Th_1 or Th_2. This formulation of t can be used to summarize the type of immune response mounted against a single-species infection or against a suite of infections.

Establishing a common immunological currency (such as summary variable t) does have costs. One difficulty with t is that it glosses over details of multi-component immune responses (such as the contribution of non-Th cells to the cytokine milieu). The systemic formulation of t also ignores the site specificity of immune responses, although t could easily be scaled down to the organ or tissue level. Finally, t is problematic in that the formulation is sensitive to details of cytokine choice and measurement. Cytokines other than IFN-γ and IL-4 might, for example, better characterize certain immune responses. Any cytokines included in the calculation of t have to be of similar order of magnitude, and all must be measured in metric (rather than antibody manufacturers') units. For all of these reasons, equation (1) may ultimately require refinement. It may even prove better to focus upon cell numbers (via flow cytometric analysis of the T-helper cell populations themselves) rather than cytokine concentrations. As a first attempt, however, t stands as a working summary of immune system action. Empirical data on optimal murine responses to infection may readily be summarized via equation (1) and then incorporated into the model described below.

An optimality model for immune responses to concomitant infection

Can the best T-helper response to concomitant infection (i.e. the response that minimizes morbidity) be predicted from the empirically-determined optimal T-helper responses to each component infection? The model is based on two principles: (a) that an optimal response to concomitant infection will minimize divergence of the joint response from each of the single-infection optima (which would predict a simple average response); and (b) that the divergence of the joint response from each local optimum should be weighted by the consequences of mounting a sub-optimal response to each of the component single infections.

The optimal value of t maximizes survival $w(t)$, as follows:

$$w(t) = \exp\left(-\sum_{i=1}^{n} \alpha_i(t-t_i)^2\right), \tag{2}$$

where n = the number of parasites/pathogens co-infecting a host; t_i = the optimal immune response to the ith parasite; and α_i = the consequences of a sub-optimal response to the ith parasite. Each t_i may be empirically determined in mouse model systems: the value of t that maximizes host survival or minimizes morbidity may be called the optimal response to infection. Alternatively, the mean value of t expressed by hosts could be construed as the optimum, with the standard deviation as a measure of the importance of mounting the correct response to a given parasite. In this way, weighting factor α_i may be estimated from data – for example, as the inverse of the variance about the empirically-determined optimum. Another possibility is that α_i could be estimated from the relationship between morbidity or mortality and a spectrum of immune responses, t. Data on infected gene-knockout mice could provide this information on the pathological consequences of sub-optimal immune responses (i.e. ones outside the spectrum of Wild-type responses). It is important to note that the magnitude of α_i is liable to vary from infection to infection. That is, the importance (to host survival and health) of getting the immune response exactly right is not the same for all parasites/pathogens. Where there is a larger 'margin for error', the weighting factor α_i would be smaller.

Solving for the optimal t for concomitant infection results in:

$$t^* = \frac{\sum \alpha_i t_i}{\sum \alpha_i}. \qquad (3)$$

The model based on weighted minimization of divergence from local optima thus predicts a weighted average response as the optimal immune response to concomitant infection: the average of all t_i weighted by the penalties for getting each response wrong. The weighted-average hypothesis may be tested against empirical data on immune responses in concomitantly-infected mice.

DATA SETS FOR TESTS OF THE MODEL

Some experiments in mouse model systems are targeted at responses to concomitant infection. Uninfected and singly-infected animals serve as controls for the concomitantly-infected animals, and researchers routinely measure cytokine concentrations, Th_1/Th_2 cell population sizes, and/or cytokine-encoding mRNA quantities, and host survival and morbidity. A single experiment may thus simultaneously generate estimates of t_i and α_i for single-species infections as well as data on the immune response against concomitant infection (t_{co}) that may be compared with t^*.

Concomitant infection by the parasites *Schistosoma mansoni* and *Toxoplasma gondii* will provide the first test of the weighted-average hypothesis. The optimal immune response against *T. gondii* infection is dominated by Th_1 cells and Th_1-associated cytokines (Denkers & Gazzinelli, 1998). The optimal response to *S. mansoni* infection, on the other hand, is Th_2 dominated (Brunet *et al.* 1998). For concomitant infection by disparate parasites such as these, the distributions of t_i will be non-overlapping. The model described above can generate formalized, quantitative (and hence readily testable) predictions about where the joint response t_{co} will fall along the spectrum of immune responses between the *Toxoplasma* and *Schistosoma* extremes. This test of the model is under way.

DISCUSSION

Can immune responses to concomitant infections be understood directly from first principles – the mechanistic understanding of single infection optimal responses – or are there emergent properties? If a simple weighted-average hypothesis (equation (3)) accurately predicts the best response to concomitant infections, then single-infection immune response data may provide sufficient information to enable us to understand the immunology of concomitantly-infected hosts. Experiments on concomitant infections are necessarily more complicated and difficult to carry out than single-species infections. It would be valuable to know whether we can learn about immune responses to concomitant infections by simply combining the results on the component single-species infections. If, on the other hand, the weighted-average hypothesis fails to predict t_{co} mounted against concomitant infections (i.e. there are emergent properties of the infection interaction), then further mathematical analysis can help to tease apart the key factors that might make the dynamics of concomitant infections differ from the sum of their component single infections. Order of infection and the intensity of each infection are particularly likely to affect immune responses to concomitant infection.

Through this optimality modelling approach, we might learn more about the complexities of the immune system, including the extent to which immune response-mediated interactions among infections might affect disease outcome, vaccine or treatment efficacy, and the like. Concomitant *Schistosoma-Toxoplasma* infections have been shown to exacerbate progression of both diseases, probably via enhanced immunopathology (Marshall *et al.* 1999). A mathematical investigation of the optimality of immune responses to concomitant *Schistosoma-Toxoplasma* infections would bring us a step towards a better understanding of emergent immunopathological outcomes of concomitant infection. Much remains to be done to develop this line of thinking, but it appears to be a promising approach to the difficult problem of concomitant infection.

ACKNOWLEDGEMENTS

My thanks go to the lab groups of E. Y. Denkers and E. J. Pearce for their many contributions to the development of these ideas. H. K. Reeve and S. E. Roberts were of great assistance in developing the optimality model. I also thank F. E. G. Cox, C. D. Harvell, and A. F. Read for encouraging me in this work. A. L. G. is a pre-doctoral fellow of the Howard Hughes Medical Institute.

REFERENCES

ALLEN, J. E. & MAIZELS, R. M. (1997). Th₁–Th₂: reliable paradigm or dangerous dogma? *Immunology Today* **18**, 387–392.

BRUNET, L. R., DUNNE, D. W. & PEARCE, E. J. (1998). Cytokine interaction and immune responses during *Schistosoma mansoni* infection. *Parasitology Today* **14**, 422–427.

CURRY, A. J., ELSE, K. J., JONES, F., BANCROFT, A., GRENCIS, R. K. & DUNNE, D. W. (1995). Evidence that cytokine-mediated immune interactions induced by *Schistosoma mansoni* alter disease outcome in mice concurrently infected with *Trichuris muris*. *Journal of Experimental Medicine* **181**, 769–774.

DENKERS, E. Y. & GAZZINELLI, R. T. (1998). Regulation and function of T-cell-mediated immunity during *Toxoplasma gondii* infection. *Clinical Microbiology Reviews* **11**, 569–588.

FISHMAN, M. A. & PERELSON, A. S. (1999). Th₁/Th₂ differentiation and cross-regulation. *Bulletin of Mathematical Biology* **61**, 403–436.

INFANTE-DUARTE, C. & KAMRADT, T. (1999). Th₁/Th₂ balance in infection. *Springer Seminars in Immunopathology* **21**, 317–338.

JANEWAY, C. A., TRAVERS, P., WALPORT, M. & CAPRA, J. D. (1999). *Immunobiology: The Immune System in Health and Disease*, New York. Elsevier Science, Ltd./Garland Publishing, Inc.

MARSHALL, A. J., BRUNET, L. R., VAN GESSEL, Y., ALCARAZ, A., BLISS, S. K., PEARCE, E. J. & DENKERS, E. Y. (1999). *Toxoplasma gondii* and *Schistosoma mansoni* synergize to promote hepatocyte dysfunction associated with high levels of plasma TNF-alpha and early death in C57BL/6 mice. *Journal of Immunology* **163**, 2089–2097.

MOREL, B. F., KALAGNANAM, J. & MOREL, P. A. (1992). Mathematical modeling of Th₁–Th₂ dynamics. In *Theoretical and Experimental Insights into Immunology*, Vol. 66 (ed. Perelson, A. S. & Weisbuch, G.), pp. 171–190. Berlin, Springer-Verlag.

MOSMANN, T. R. & SAD, S. (1996). The expanding universe of T-cell subsets: Th₁, Th₂ and more. *Immunology Today* **17**, 138–146.

PARKER, G. A. & MAYNARD-SMITH, J. (1990). Optimality theory in evolutionary biology. *Nature* **348**, 27–33.

ROMAGNANI, S. (1996). Th₁ and Th₂ in human diseases. *Clinical Immunology and Immunopathology* **80**, 225–235.

SCHWEITZER, A. N., SWINTON, J. & ANDERSON, R. M. (1993). Dynamic interaction between *Leishmania* infection in mice and Th₁-type CD4+ T-cells: complexity in outcome without a requirement for Th₂-type responses. *Parasite Immunology* **15**, 85–99.

SHER, A. & COFFMAN, R. L. (1992). Regulation of immunity to parasites by T cells and T cell-derived cytokines. *Annual Review of Immunology* **10**, 385–409.

Parasitic diseases and immunodeficiencies

P. AMBROISE-THOMAS

Laboratoire Interactions Cellulaires Parasite-Hôte (ICPH), Faculté de Médecine, Université Joseph Fournier-Grenoble, E.R., CNRS 2014, 38043 Grenoble, France

SUMMARY

In the last two decades, major immunodeficiency syndromes have strongly influenced medical parasitology. Some animal parasitoses, once unknown in human medicine, have become zoonotic and sometimes anthroponotic. In other cases, the clinical evolution of human parasitoses has been severely aggravated and/or modified in immunodeficient patients especially in toxoplasmosis, cryptosporidiosis, leishmaniasis, strongyloidiasis and scabies. The parasites implicated are varied (protozoa, helminths and even Acaridae) but have in common the capacity to reproduce in or on the human host. These immunodeficiency syndromes are often related to AIDS but other major immunodepressions, such as post-therapeutically in organ transplantation, may also be responsible and raise difficult problems for prevention. The immunological mechanisms involved are not always well understood. In addition, genetic predisposition factors, gradually becoming better-understood in parasites and man, complete and complicate our understanding of the immunological mechanisms.

Key words: Opportunistic parasitoses, immunodeficiencies, AIDS, organ transplants.

INTRODUCTION

The advent of major immunodeficiency syndromes has opened a new and important chapter in medical parasitology, that of 'opportunistic' parasitoses and, to some extent, of 'emergent' parasitoses. Among these immunodeficiencies, AIDS obviously ranks first. It should be noted that 3 of the 12 case-defining opportunistic infections for AIDS (AIDS group IV, according to the Centers for Diseases Control classification) are parasitic diseases (toxoplasmosis, cryptosporidiosis, and isosporosis).

Besides AIDS, other immunodeficiency syndromes of pathological origin such as cancers and leukaemias, or induced by immunosuppressive therapy (intensive corticotherapy, organ transplants) may also strongly modify the course and presentation as well as the epidemiology of various human parasitoses. In some cases, host specificity and thus inter-species barriers can decrease or disappear in the immunodeficient host: animal parasites may infect humans and become zoonotic then anthroponotic as described by Charles Nicolle in *Destin des maladies infectieuses* (Ambroise-Thomas & Grillot, 1995).

Several parasitic diseases illustrate the various case-scenarios mentioned above. All are caused by parasites that involve humans in their asexual and/or sexual cycles. Most often, they are endocellular protozoans belonging to different genera or species (*Toxoplasma gondii, Cryptosporidium parvum* and *Cyclospora cayetanensis, Isospora belli* and *Sarcocystis* spp., *Microsporidia sensu lato, Leishmania* spp.,

Trypanosoma cruzi), but also nematodes and Acaridae (*Strongyloides stercoralis* and *Sarcoptes scabiei*).

Within this review, we will (1) mention briefly the main clinical and epidemiological modifications linked to major immunodeficiencies in these parasitic diseases; (2) study the causal mechanisms according to the type of immunodeficiency; and (3) describe general phenomena caused by the opportunistic parasite.

OPPORTUNISTIC PARASITOSES IN IMMUNOCOMPROMISED PATIENTS

Toxoplasmosis

Toxoplasmosis is the most frequent and best known of the various opportunistic parasitic diseases. In immunocompetent patients, infection is normally asymptomatic or unremarkable. Acquired toxoplasmosis normally becomes latent and parasite persistence in the form of cysts leads to the build-up of a protective immunity. In contrast, congenital toxoplasmosis may cause intra-uterine death, neonatal growth retardation, mental retardation, ocular defects and blindness. Similar degrees of severity are found in immunocompromised patients (Ambroise-Thomas & Pelloux, 1993), notably in AIDS patients with cerebral, often life-threatening, manifestations (Leport & Remington, 1992). Multiple visceral localisations (lung, liver, kidney) are also observed (Rabaud *et al.* 1996) while retinochoroiditis is less common than in congenital toxoplasmosis, for reasons that are not well understood. These cases of toxoplasmosis in AIDS patients almost always result from the reactivation of an old and unrecognised

E-mail: PAmbroise-Thomas@chu-grenoble.fr
Tel: (33)4 76 76 55 72. Fax: (33)4 76 76 56 60.

chronic toxoplasmosis, when the CD4+ lymphocyte count is under 100 μl. The same mechanism may occur in life-threatening toxoplasmosis following immunodepressant therapy following organ transplantation (Couvreur *et al.* 1992; Derouin *et al.* 1992; Gallino *et al.* 1996). Occasionally, infection may not be caused by a reactivation of latent toxoplasmosis (as in AIDS patients) but rather from a primary infection resulting from an organ or tissue transplantation and aggravated by immunodepressant agents (Mayes *et al.* 1995). Without the currently used preventive measures, this risk would reach approximately 50% of heart transplant cases, 20% of liver transplants, and only 2% of kidney or bone marrow transplants (Joynson, 1999), which well illustrates the preferential sites of *Toxoplasma gondii* cysts or tachyzoites.

Cryptosporidium *and* Cyclospora *infections*

In AIDS patients the *Cryptosporidium parvum* and *Cyclospora cayetanensis* parasite load is particularly high. The number of parasites increases through merozoite auto-reinfections emerging from type 1 schizonts or from zygotes (Bonnin *et al.* 1998). Consequently clinical manifestations are particularly severe, with abundant diarrhoea (up to 17 l/day) leading to serious nutritional and hydroelectric consequences. The most prominent feature of cryptosporidiosis in AIDS is the risk of (possibly haematogenous) dissemination with secondary multiple localisation in the liver, pancreas, stomach, and respiratory tract (Pinel *et al.* 1998). Nevertheless besides serious complications in patients with severe immunodepression (CD4+ < 50 μl), less serious (paucisymptomatic) manifestations occur in patients with CD4+ counts > 180–200 μl. Lastly, as well as in AIDS, severe disseminated cryptosporidiosis may occur in other cases of immunodepression, particularly those due to cancer, transplants, dialyses (Gascon, Zabala & Iglesias, 1998) as well as in patients presenting with congenital IgG or IgA deficiencies.

Isosporosis and sarcocystosis

Of all the species belonging to the genus *Isospora*, only *Isospora belli* may be considered as a truly opportunistic parasite. Very rarely found in immunocompetent subjects (less than 0·2% in Southeast Asian refugees) it is however very frequent in AIDS patients, especially in tropical countries (up to 85% of AIDS patients in Haiti) and causes chronic diarrhoea leading to serious absorption problems (Dei-Cas, 1994). By contrast *Sarcocystis bovihominis* and *S. suihominis*, which cause moderate digestive problems in immunocompetent subjects, do not present any particular pathogenicity in patients presenting with AIDS or other immunodeficiencies.

Microsporidioses

Microsporidia are found world-wide and infect vertebrates and invertebrates alike. Approximately 100 different genera and 1000 species are known; in man only 12 species belonging to 7 genera have been described: *Enterocytozoon, Encephalitozoon, Nosema, Microsporidium, Trachipleistophora, Vittaforma* and *Brachiola*.

Until 1985 microsporidioses were virtually unknown in human pathology since they were only rarely observed and, for the most part, diagnosis was questionable. It was with the emergence of AIDS that this parasitosis became fully recognised. In immunocompetent patients, microsporidioses are mainly asymptomatic or paucisymptomatic, only rarely causing diarrhoea in travellers or exceptionally ocular or cerebral effects. In AIDS patients, these opportunistic parasitoses are normally severe (Bryan, 1995; VanGool & Dankert, 1995). Localisation and clinical presentation vary according to the parasite species: intestinal or hepatic (*Encephalitozoon bieneusi, Encephalitozoon intestinalis*), disseminated/systemic with polyvisceral localisation (*Encephalitozoon hellem, E. cuniculi* and *Vittaforma corneae*), ocular localisation has also been reported for several species including *Microsporidium ceylonensis* and *M. africanum* (Schwartz *et al.* 1996).

However, even in immunodepressed patients, microsporidiosis could be clinically mild. Unlike toxoplasmosis or cryptosporidiosis, CD4+ lymphocyte counts have no prognostic value in AIDS. Besides AIDS, other severe immuno-depressions before organ transplant (renal, allogenic marrow, heart-lung, liver) can cause severe microsporidioses (Sax *et al.* 1995; Rabodonirina *et al.* 1996; Kelkar *et al.* 1997; Gumbo *et al.* 1999; Metge *et al.* 2000).

Leishmaniasis and Chagas Disease

In endemic areas, leishmaniasis is the most frequent opportunistic parasitosis in AIDS after toxoplasmosis and cryptosporidiosis (Desjeux, 1996; Dedet & Pratlong, 2000). In fact AIDS multiplies the risk of visceral leishmaniasis 100 to 1000 fold and increases its severity, particularly in *L. infantum* infections in south-western Europe (WHO, 1999). Visceral leishmaniasis also increases immunodepression and accelerates viral replication and transition from HIV infection to overt AIDS, at the same time increasing the risk of other opportunistic infections.

Among the AIDS cases reported by the WHO, 90·4% of patients had less than 200 CD4+/μl. Even in AIDS patients, asymptomatic leishmaniasis may be observed in approximately 10% of the cases along with the classical forms comparable to those in immunocompetent patients combining fever, hepatosplenomegaly and pancytopenia (82·4% of the cases)

(Marlier *et al.* 1999). Atypical forms have also been described. These are either paucisymptomatic, corresponding most often to an early diagnosis, or by contrast they feature localisations which are unusual in the immunocompetent host: intestinal localisations with erosive or more rarely tumoral lesions, cutaneous lesions which may be indicative of concomitant visceral dissemination, pleuro-pulmonary (rare), laryngeal or neuromeningeal forms (extremely rare).

Mortality due to leishmaniasis is high (10–20 % during the first episode), nevertheless newer combination antiviral drugs (tri-therapy) are likely to modify the course of the disease by improving the immune response and thus increasing the efficacy of anti-*Leishmania* treatments (Behre *et al.* 1999).

No change in frequency and presentation has been reported for African trypanosomiasis in HIV patients, while reactivation of Chagas disease is observed in patients dually infected with HIV and *Trypanosoma cruzi* (Dedet & Pratlong, 2000).

Strongyloidiasis

In immunocompromised patients, *Strongyloides stercoralis* may cause hyperinfections due to a considerable increase of the auto-reinfection cycle with a dissemination of L_3 larvae throughout the whole host, and of adult females, eggs, and L_1 larvae outside of the digestive tract. Clinically, digestive symptoms are particularly important (severe diarrhoea, emesis, abdominal pain) with various visceral localisations: intestinal (ileum necrosis, ulcerative colitis, peritonitis), pulmonary, often over-infected with Gram negative bacteria (Gulletta *et al.* 1998), or neurological. These disseminated malignant strongyloidiases are often lethal. Observed for the first time in mentally retarded children with congenital immunodeficiencies (Yoeli *et al.* 1963), they are mainly caused by intensive and long-term corticotherapy when treating auto-immune diseases, lymphomas etc. The mechanism of action on the parasites is not well understood, but corticotherapy may accelerate extra-intestinal transformation of L_1 into L_3 larvae. This probably explains only part of the phenomenon because disseminated strongyloidiasis also occurs after reactivation of chronic strongyloidiasis due to the immunosuppressive therapy in transplant recipients, particularly in kidney transplantation (De Vault *et al.* 1990). This risk is frequent and justifies screening for strongyloidiasis in all patients from endemic strongyloidiasis areas and, if necessary, to cure it before organ transplantation. In some very exceptional cases, the graft itself could be the source of contamination (Palau & Pankey, 1997).

The influence of AIDS is still controversial. Nevertheless, disseminated malignant strongyloidiases have been observed more frequently in patients infected by HTLV1 which, for unknown reasons, seems to facilitate these disseminated strongyloidiasis more than HIV (Neisson-Vernant & Edouard, 1990).

Scabies

In immunodeficient patients and particularly in AIDS patients, scabies has a special clinical picture, with widespread or localised hyperkeratotic generally non-pruritic skin lesions. In some cases these crusted lesions are pruritic and resemble Darier's disease or psoriasis (Schlesinger, Oerlich & Tyring, 1994). This 'Norwegian' scabies is extremely contagious and undiagnosed cases can be a source of nosocomial infections (Guggisberg *et al.* 1998). Apart from AIDS, other immunodeficiencies, particularly those expected in old patients, can also induce this type of complicated scabies.

Other parasitic diseases and opportunistic mycoses

This already long and varied list could be completed with other parasitic diseases which are rarely truly opportunistic, but the clinical symptoms of which could be modified during major immunodeficiencies of pathological or therapeutic origin. This is the case of *Giardia duodenalis* infection (which is worsened by IgA deficiency), of *Blastocystis hominis* infection (which could be the cause of severe diarrhoea in AIDS patients), or of *Babesia divergens* or *B. microti* infections (which only occur in splenectomised patients).

One could also add numerous opportunistic mycoses: pneumocytosis, crypotococcosis, invasive aspergillosis, oro-pharyngeal or disseminated candidiasis. Although in several countries, including France, mycology and parasitology are linked, we will not consider these fungal infections within this review.

MECHANISMS

The various types of immunodeficiencies are, of course, the first and main mechanism of outbreaks of opportunistic parasitoses. Nevertheless, these immunodeficiencies, whatever their type and severity, are not the only triggering mechanism. Parasitic genotypic and phenotypic characteristics also play a major role, as well as human genetic factors conditioning sensitivity to some parasitic infections.

The various immunodeficiencies and their consequences

Specific anti-parasitic immunity is produced by numerous and complicated mechanisms mainly implicating three specific cellular populations, phagocytes, and both B and T lymphocytes. For each parasitosis, it is usually possible to demonstrate that one of these factors plays an essential role. Conversely, it has been shown that a given immunodepression would promote a specific type of opportunistic parasitosis. Nevertheless, this is not universal since anti-parasitic immunity almost always results from a combination of various specific and non-specific factors.

Immunodeficiencies act in a predictable way, notwithstanding the diversity of opportunistic parasitoses. They do not really increase the intrinsic pathogenicity of parasites (except maybe in the case of cryptosporidiosis), but they promote their replication, either at the onset (strongyloidiasis for example), or after reactivation of a quiescent infection (toxoplasmosis).

In immunocompetent patients, toxoplasmosis becomes and remains latent due to several complex immunological mechanisms, some of which are not well understood. Humoral immunity plays a role since antibodies specific to the major surface protein of tachyzoites (SAG1) inhibit both the attachment of the parasite to the host cell receptors and then its infection. Nevertheless cellular immunity remains the main component of this immunity. CD4+ lymphocytes play an important but still controversial part even if it is their count, in AIDS patients, which permits the definition of the alert threshold (< 100 CD4+/μl). But CD4+ lymphocytes act synergistically with CD8+ lymphocytes which are the key elements of anti-toxoplasmic immuno-protection (Kasper & Buzoni-Gatel, 1998). In the course of AIDS, severe toxoplasmosis only develops after several years, specifically when CD8+ lymphocyte counts begin to decrease. After stimulation by toxoplasmic antigens and with IL-12 acting as a mediator, these cells secrete γ Interferon (IFN-γ) which acts simultaneously with α Tumour Necrosis Factor (TNF-α) in the production of NO. Other cytokines (IL-4, IL-6, IL-7, IL-15) may also be implicated in these protective mechanisms involving other cell populations (dendritic cells). Besides this specific immunity, other phenomena are essential for the protection against toxoplasmosis such as natural killer cells (NK) which, when activated by IL-12 secreted by activated macrophages, are one of the main sources of IFN-γ. HIV-1 also has a direct effect, by reducing the phagocytic and parasiticidal function of macrophages (Biggs *et al.* 1995).

These mechanisms, independent of CD4+ or CD8+ lymphocyte action, may account for the very paradoxical case of malaria (Butcher, 1992). Indeed, *Plasmodium falciparum* is a sporozoan like *Toxoplasma gondii* and *Cryptosporidium parvum*. Nevertheless, malaria is seldom worsened by AIDS, in contrast to toxoplasmosis and cryptosporidiosis (Leaver, Haile & Watters, 1990). This is all the more unexpected since other viral co-infections induce very severe complications in the course of malaria

(*Plasmodium falciparum* and Epstein-Barr virus co-infections in Burkitt's lymphoma). In fact, apart from humoral and mainly cellular immunity (CD4+ Th$_1$), protection against malaria is due to macrophages which, after having been activated by plasmodial antigens, produce H$_2$O$_2$, NO, and TNF-α (which is a cytotoxic factor acting on *P. falciparum*). This may explain the lack of evidence for an interaction between HIV infection and malaria, although the risk of malaria during pregnancy was higher in HIV positive than in HIV negative women in Malawi (Verhoeff *et al.* 1999). Nevertheless most studies did not show this interaction between HIV and malaria, but they had several methodological limitations. Conversely, recent data shown that *P. falciparum* malaria may increase the HIV-1 viral load (Hoffmann *et al.* 1999). This may at least partially account for the higher number of AIDS cases in HIV infected patients, in tropical or subtropical countries more than in countries free of malaria (Chandramohan & Greenwood, 1998).

For cryptosporidiosis, circulating antibodies are partially responsible for the control of the infection, as shown by the severity of this parasitosis during congenital IgA and IgG deficiencies. Nevertheless, this is also the case for other opportunistic protozoan infections; cellular immunity is the most effective protective mechanism, particularly through CD4+ T lymphocytes. Below a determined value (50 CD4+/μl) the CD4+ count is also one of the main alarm signals concerning the risk of severe cryptosporidiosis during AIDS. In immunocompetent patients, CD4+ T lymphocytes in Peyer's patches or in the intestinal lymph nodes are activated by parasite antigens trapped by cells presenting antigens. These lymphocytes then migrate to the intestinal mucosae where they synthesise various cytokines, particularly IFN-γ which acts directly on enterocytes and can activate macrophages and lymphocytes of the mucosae. Apart from IFN-γ, other cytokines such as TNF-α and IL-8, along with chemokines, play an important part in the protective immunity against *Cryptosporidium parvum*. The role of CD8+ T lymphocytes is more controversial and is probably less important than in toxoplasmosis. In cryptosporidiosis, a decrease of humoral immunity may also cause important consequences with a drop in circulating specific antibody titres which may neutralise a cryptosporidial toxin with a pathophysiological role.

The problem is even more complicated in the leishmaniases. Indeed, as shown in several studies, there is a vicious circle in which visceral leishmaniasis and AIDS mutually reinforce their effects. These interactions are probably partially explained by the fact that *Leishmania* and HIV can infect and multiply in the same type of cells (macrophages). Visceral leishmaniases increase the HIV transactivation in chronically infected monocytes and in

CD4+ T lymphocytes. These parasitoses decrease the cytotoxic HIV-specific effect in CD8+ T and promote the transition from HIV-infection to AIDS-disease, with an increased risk of other opportunistic infections, particularly by endocellular bacteriae (mycobacteriae). Conversely, HIV deactivates the macrophage functions, depresses the activity of Th_1 cells, enhances the activity of Th_2 cells, and inhibits phagosome–lysosome fusion (Wolday *et al.* 1999). Thus, HIV can enhance intracellular growth then the dissemination of *Leishmania*, and considerably increase the severity of visceral leishmaniasis.

The exact mechanisms by which malignant disseminated strongyloidiases are caused during various types of immunodeficiencies are not well understood. In particular, the role of AIDS is still controversial and so are the reasons why HTLV-1 may facilitate the *Strongyloides stercoralis* multiplication and dissemination more than HIV (Neisson-Vernant & Edouard, 1990). Apart from various immunological factors, some direct mechanisms may also cause severe disseminated strongyloidiasis, as may be the case during massive corticotherapy with a direct influence of corticoids on the parasite.

Finally, in Norwegian scabies, the data are rare, often incomplete, and mainly correspond to very variable immunological contexts, including physiological, gerontological and mild immunodeficiencies. We know practically nothing about the immune mechanisms which permit proliferation of an ectoparasite such as *Sarcoptes scabiei* and which considerably increase the severity and the transmissibility of scabies.

Genetic mechanisms

Immunological mechanisms have been pointed out as having a key role in triggering opportunistic parasitoses, especially since it is possible to determine risk circumstances in some cases (AIDS for example) and even to assess that risk with an alert threshold (CD4+ lymphocyte count).

Even if they are very important, these immunological causes are not the only ones. In the case of toxoplasmosis in patients with a latent toxoplasmosis (positive toxoplasmic serology), the onset of cerebral toxoplasmosis in AIDS patients may be determined by a CD4+ lymphocyte count less than 100 CD4+/ μl. Nevertheless, this alert threshold has a relative value and should not be wrongly interpreted since very severe toxoplasmic reactivations are observed only in 30% of AIDS patients with a positive toxoplasmic serology, even with a CD4+ count less than 100 μl. These differences have been tentatively explained by pathogenicity specific to some strains of *Toxoplasma gondii* (Gross *et al.* 1997). Indeed, genomic virulence markers have been identified (Guo, Gross & Johnson, 1997) and for some, located on

chromosome VIII (Howe, Summers & Sibley, 1996), but these genotype features may be modified by various mechanisms (Frenkel & Ambroise-Thomas, 1997). Isoenzyme studies have completed these data by identifying particular zymodemes which, associated with genotypes, allow better characterisation of the strains (Cristina *et al.* 1995). Some profiles are particularly associated with severe toxoplasmoses, but this relationship is especially true for experimental infections in the murine model. The situation is not as clear concerning human opportunistic toxoplasmoses for which neither the features of the parasitic strain nor the intensity of immunodepression can explain the severity or the mildness of the disease. The host's genetic predisposition features also play a role (may be essential?) (Brown *et al.* 1995). These features are gradually becoming better understood and are related to genes belonging to the class II HLA group (Mack *et al.* 1999) or to genes independent of the Major Histocompatibility Complex (Suzuki *et al.* 1996).

In cryptosporidiosis as well as in toxoplasmosis, features directly linked to the parasite may determine the severity of the disease. Indeed, thanks to the techniques of molecular biology, various genotypes able to infect man have been identified in *Cryptosporidium parvum*. Four of these genotypes are found in AIDS patients (Pieniazek *et al.* 1999). Three of these are also found in animals (zoonotic cryptosporidiosis) whereas the type 1 genotype is strictly human (anthroponotic cryptosporidiosis) and could be linked to the most severe human infections. The fact that this "human" genotype exists shows that *C. parvum* has gone through a recent biological evolution, with a loss or modification of its host specificity allowing it to reproduce only in the human host. This phenomenon probably comes from parasite mutations and, in humans, from selection due to adaptative pressure, may be facilitated by major immunodeficiencies. The emergence of human cryptosporidiosis has come about after the appearance of AIDS and the most active immunosuppressant treatments. By neutralising the immunological mechanisms partially responsible for the species barrier, these immunodepressions have allowed the parasite to reproduce intensively in humans, notably due to a sexual cycle of auto-reinfestation. This is what may have promoted the emergence and the reproduction of variants specifically (?) adapted to parasitism in man. Of course, the mechanisms of this emergence are not only immunological and the great number of human cryptosporidiosis cases reported during various epidemics (400000 cases in Wisconsin, in the USA, in 1993) must also have promoted this evolution. Finally, among the factors of clinical severity, host related genetic factors also play a part. This was demonstrated by experimental infections in the murine model (Davami, Bancroft &

McDonald, 1997) and very probably takes place in man, even if the genes responsible have not been identified yet.

The same genetic predisposition factors, in the parasite and in the host, also play a role in the leishmaniases in immunodepressed patients. Thus, isoenzymatic analysis of more than 250 isolates of *Leishmania infantum* has identified 18 different parasitic zymodemes, 10 of which are found only in HIV co-infections. Thus genetic predisposition is probably a common mechanism in the onset of opportunistic parasitoses, even if it has not been proved in all the cases, especially for opportunistic parasitoses with helminths (strongyloidiasis) or Acaridae (Norwegian scabies).

CONCLUSION

The great discoveries of the 19th and 20th centuries have lead to great progress in parasitology. They may have also led to an oversimplified view of the parasitic phenomenon, with too much distinction between human and animal parasites, between harmless and pathogenic or very pathogenic parasites. In fact, things are not as clear-cut. Parasitism is an evolutive phenomenon. This evolution is ongoing and in the last two decades, it has experienced a sharp acceleration with the onset of major immunodeficiencies, in human diseases. Thus, in a very short time-span, animal parasites lost their host specificity and became transmissible to man, while others evolved to become exclusively specific to man, while still others demonstrated a pathogenicity up until then unknown. All the available techniques in immunology and molecular genetics have been used to explain this evolution. Other methods will certainly appear in the future and will allow us to understand these phenomena better. At this point, current results have the merit of reminding us how fluctuating the fate of parasitic diseases can be and underlining the role played, besides the parasite and the host, by co-infective microorganisms and especially viruses. Mono-infestations of course do not exist in nature, and parasitism finally seems to be the end-product of multiple interactions between the host and the complex microbial, viral, parasitic and fungal influences.

REFERENCES

AMBROISE-THOMAS, P. & GRILLOT, R. (1995). Une étape très actuelle du destin des maladies infectieuses: les infections opportunistes parasitaires et fongiques. *Bulletin de l'Académie Nationale de Médecine* **179**, 789–803.

AMBROISE-THOMAS, P. & PELLOUX, H. (1993). Toxoplasmosis. Congenital and in immunocompromised patients. A parallel. *Parasitology Today* **9**, 61–63.

BERHE, N., WOLDAY, D., HAILU, A. *et al.* (1999). HIV viral load and response to anti-leishmanial chemotherapy in co-infected patients. *AIDS* **13**, 1921–1925.

BIGGS, B.-A., HEWISH, M., KENT, S. *et al.* (1995). HIV-1 infection of human macrophages impairs phagocytosis and killing of *Toxoplasma gondii*. *Journal of Immunology* **154**, 6132–6139.

BONNIN, A., DUBREMETZ, J. F., LOPEZ, J. *et al.* (1998). Infections à Cryptosporidies et à *Cyclospora*. *Encyclopédie Médico-Chirurgicale, Maladies Infectieuses* **501 A10**, 1–9.

BROWN, C. R., HUNTER, C. A., ESTES, R. G. *et al.* (1995). Definitive identification of a gene that confers resistance against *Toxoplasma* cysts burden and encephalitis. *Immunology* **845**, 419–428.

BRYAN, R. T. (1995). Microsporidiosis as an AIDS-related opportunistic infection. *Clinical Infectious Diseases* **2**, S62–S65.

BUTCHER, G. A. (1992). HIV and malaria: a lesson in immunology? *Parasitology Today* **8**, 307–311.

CHANDRAMOHAN, D. & GREENWOOD, B. M. (1998). Is there an interaction between human immunodeficiency virus and *Plasmodium falciparum*? *International Journal of Epidemiology* **27**, 296–301.

COUVREUR, J., TOURNIER, G., SADET-FRISMAND, A. *et al.* (1992). Transplantation cardiaque ou cardio-pulmonaire et toxoplasmose. *La Presse Médicale* **2**, 1569–1574.

CRISTINA, N., DARDE, M. L., BOUDIN, C. *et al.* (1995). DNA fingerprinting method for individual characterization of *Toxoplasma gondii* strains: combination with isoenzymatic characters for determination of linkage groups. *Parasitology Research* **8**, 32–37.

DAVAMI, M. H., BANCROFT, G. J. & MCDONALD, V. (1997). *Cryptosporidium parvum* infection in major histocompatibility complex congenic mice: variation in susceptibility and the role of T-cell cytokine responses. *Parasitology Research* **8**, 257–263.

DEDET, J. P. & PRATLONG, F. (2000). *Leishmania*, *Trypanosoma* and monoxenous trypanosomatids as emerging opportunistic agents. *Journal of Eukaryotic Microbiology* **47**, 37–39.

DEI-CAS, E. (1994). Infections à microsporidies, *Isospora* et *Sarcocystis*. *Encyclopédie Médico-Chirurgicale, Maladies Infectieuses* **503 A10**, 1–6.

DEROUIN, F., DEVERGIE, A., AUBER, P. *et al.* (1992). Toxoplasmosis in bone-marrow-transplant recipients. Report of seven cases and review. *Clinical Infectious Diseases* **15**, 267–270.

DESJEUX, P. (1966). Leishmaniasis. Public health aspects and control. *Clinical Dermatology* **14**, 417–423.

DE VAULT, Jr G. A., KING, J. W., ROHR, M. S. *et al.* (1990). Opportunistic infection with *Strongyloides stercoralis* in renal transplantation. *Review of Infectious Diseases*. **12**, 653–671.

FRENKEL, J. K. & AMBROISE-THOMAS, P. (1997). Genomic drift of *Toxoplasma gondii*. *Parasitology Research* **83**, 1–5.

GALLINO, A., MAGGIORINI, M., KIOWSKI, W. *et al.* (1996). Toxoplasmosis in heart transplant recipients. *European Journal of Clinical Microbiology and Infectious Diseases* **15**, 389–393.

GASCON, A., ZABALA, S. & IGLESIAS, E. (1998). Cryptosporidiosis in haemodialysis patient with depressed CD4+ T cell count: successful treatment with azithromycin. *Nephrology Dialysis and Transplantation* **13**, 2932–2933.

GROSS, U., KEMPF, M. C., SEEBER, F. *et al.* (1997). Reactivation of chronic toxoplasmosis: is there a link to strain-specific differences in the parasite? *Behring Institute Mitt.*, **99**, 97–106.

GUGGISBERG, D., DE VIRAGH, P. A., CONSTANTIN, C. *et al.* (1998). Norwegian scabies in a patient with acquired immunodeficiency syndrome. *Dermatology* **197**, 306–308.

GULLETTA, M., CHATEL, G., PAVIA, M. *et al.* (1998). AIDS and strongyloidiasis. *International STD AIDS* **9**, 427–429.

GUMBO, T., HOBBS, R. E., CARLYN, C. *et al.* (1999). *Microsporidia* infection in transplant recipients. *Transplantation* **67**, 482–484.

GUO, Z. G., GROSS, U. & JOHNSON, A. M. (1997). *Toxoplasma gondii* virulence markers identified by random amplified polymeric DNA polymerase chain reaction. *Parasitology Research* **83**, 458–463.

HOFFMAN, I. F., JERE, C. S., TAYLOR, T. E. *et al.* (1999). The effect of *Plasmodium falciparum* malaria on HIV-1 RNA blood plasma concentration. *AIDS* **13**, 487–494.

HOWE, D. K., SUMMERS, B. C. & SIBLEY, L. D. (1996). Acute virulence in mice is associated with makers on chromosome VIII in *Toxoplasma gondii*. *Infection and Immunity* **64**, 5193–5198.

JOYNSON, D. H. M. (1999). Emerging parasitic infections in man. *Infectious Disease Review Medecine Veterinarian Environmental* **1**, 131–134.

KASPER, L. H. & BUZONI-GATEL, D. (1998). Some opportunistic parasitic infection in AIDS: candidiasis, pneumocystosis, cryptosporidiosis, toxoplasmosis. *Parasitology Today* **14**, 150–156.

KELKAR, R., SASTRY, P. S., KULKARNI, S. S. *et al.* (1997). Pulmonary microsporidial infection in a patient with CML undergoing allogenic marrow transplantation. *Bone Marrow Transplantation* **19**, 179–182.

LEAVER, R. J., HAILE, Z. & WATTERS, D. A. J. (1990). HIV and cerebral malaria. *Transactions of the Royal Society of Tropical Medicine and Hygiene* **84**, 201.

LEPORT, C. & REMINGTON, J. S. (1992). Toxoplasmose au cours du SIDA. *La Presse Médicale* **21**, 1165–1171.

MACK, D. G., JOHNSON, J. J., ROBERTS, F. *et al.* (1999). HLA-class II genes modify outcome of *Toxoplasma gondii* infection. *International Journal for Parasitology* **29**, 1351–1358.

MARLIER, S., MENARD, G., GISSEROT, O. *et al.* (1999). Leishmaniose et virus de l'immunodéficience humaine: une co-infection en émergence? *Médecine Tropicale* **59**, 193–200.

MAYES, J. T., O'CONNOR, B. J., AVERY, R. *et al.* (1995). Transmission of *Toxoplasma gondii* infection by liver transplantation. *Clinical Infectious Diseases* **21**, 511–515.

METGE, S., TRAN VAN NHIEU, J., DAHMANE, D. *et al.* (2000). A case of *Enterocytozoon bieneusi* infection in an HIV-negative renal transplant recipient. *European Journal of Clinical Microbiology and Infectious Diseases* **19**, 221–223.

NEISSON-VERNANT, C. & EDOUARD, A. (1990). Strongyloïdose maligne et virus HTLV-1. *Revue du Praticien* **40**, 217–2128.

PALAU, L. A. & PANKEY, G. A. (1997). *Strongyloides* hyperinfection in a renal transplant recipient receiving cyclosporine: possible *Strongyloides stercoralis* transmission by kidney transplant. *American Journal of Tropical Medicine and Hygiene* **57**, 413–415.

PIENIAZEK, N. J., BORNAY-LLINARES, F. J., SLEMENDA, S. B. *et al.* (1999). New *Cryptosporidium* genotypes in HIV-infected persons. *Emerging Infectious Diseases* **5**, 444–449.

PINEL, C., LECLERCQ, P., PICOT, S. *et al.* (1998). Cryptosporidiose pulmonaire asymptomatique chez un patients sidéen profondément immunodéprimé. *La Presse Médicale* **27**, 261.

RABAUD, C., MAY, T., LUCET, J. C. *et al.* (1996). Pulmonary toxoplasmosis in patients infected with human immunodeficiency virus: a French national survey. *Clinical Infectious Diseases* **23**, 1249–1254.

RABODONIRINA, M., BERTOCCHI, M., DESPORTES-LIVAGE, I. *et al.* (1996). *Enterocytozoon bieneusi* as a cause of chronic diarrhea in a heart-lung transplant recipient who was sero-negative for human immunodeficiency virus. *Clinical Infectious Diseases* **23**, 114–117.

SAX, P. E., RICH, J. D., PIECAK, W. S. *et al.* (1995). Intestinal microsporidiosis occuring in transplant patients. *Transplantation* **60**, 617–618.

SCHLESINGER, I., OELRICH, D. M. & TYRING, S. K. (1994). Crusted (Norwegian) scabies in patients with AIDS: the range of clinical presentation. *South Medical Journal* **87**, 352–356.

SCHWARTZ, D. A., SOBOTTKA, I., LEITCH, G. J. *et al.* (1996). Pathology of microsporidiosis: emerging parasitic infections in patients with acquired immunodeficiency syndrome. *Archives of Pathology and Laboratory Medicine* **120**, 173–188.

SUZUKI, Y., WONG, S. Y., GRUMET, F. C., *et al.* (1996). Evidence for genetic regulation of susceptibility to toxoplasmic encephalitis in AIDS patients. *Journal of Infectious Diseases* **173**, 265–268.

VAN GOOL, T. & DANKERT, J. (1995). Human microsporidiosis: clinical, diagnosis and therapeutic aspects of an increasing infection. *Journal of Clinical Microbiology and Infection* **1**, 75–85.

VERHOEFF, F. H., BRABIN, B. J., HART, C. A. *et al.* (1999). Increased prevalence of malaria in HIV-infected pregnant women and its implications for malaria control. *Tropical Medicine and International Health* **4**, 5–12.

WHO (1999). *Leishmania*/HIV co-infection, south-western Europe, 1990–1998. *Weekly Epidemiological Record* **74**, 365–376.

WOLDAY, D., BERHE, N., AKUFFO, H. & BRITTON, S. (1999). *Leishmania* – HIV interaction: immunopathogenic mechanisms. *Parasitology Today* **15**, 182–187.

YOELI, M., MOST, H., BERMAN, H. H. *et al.* (1963). The problem of strongyloidiasis among the mentally retarded in institutions. *Transactions of the Royal Society of Tropical Medicine and Hygiene* **57**, 336–345.

Multiple helminth infections in children: impact and control

L. J. DRAKE[1]* *and* D. A. P. BUNDY[2]

[1] *Scientific Coordinating Centre for the Partnership for Child Development, Wellcome Trust Centre for the Epidemiology of Infectious Disease, University of Oxford, South Parks Road, Oxford OXI 3PS*
[2] *The Human Development Network, The World Bank, 1818 H Street, Washington DC 20433, USA*

SUMMARY

Parasitic worm infections are amongst the most widespread of all chronic human infections. It is estimated that there are more than 3 billion infections in the world today. In many low income countries it is often more common to be infected than not to be. Indeed, a child growing up in an endemic community can expect be infected soon after weaning, and to be infected and constantly reinfected for the rest of her or his life. Infection is most common amongst the poorest and most disadvantaged communities, and is typically most intense in children of school going age. As the risk of morbidity is directly related to intensity of infection, it follows that children are the most at risk from the morbid effects of disease. Multiparasite infections are also common in such communities and there is evidence that individuals harbouring such infections may suffer exacerbated morbidity, making children even more vulnerable. Thus, these infections pose a serious threat to the health and development of children in low income countries. For many years, the need to control these infections has lain uncontested, and with the advent of broad-spectrum anthelminthic drugs that are cheap, safe and simple to deliver, control has at last become a viable option for many communities. Furthermore, there is now increased emphasis being placed on a multispecies approach as a cost-effective mechanism to control the morbidity of virtually all the major helminthic infections of humans.

Key words: helminth, multiple infections, school-age child.

THE MAJOR HELMINTH INFECTIONS AFFECTING CHILDREN

There are approximately 20 major helminth infections of humans that all, to some extent, have public health significance (for review refer to Warren, Bundy & Anderson, 1993), but amongst the most common of all human infections are the geohelminthiases. Recent global estimates indicate that more than one quarter of the world's population is infected with one or more of the most common of these parasites: the roundworm, *Ascaris lumbricoides*; the hookworms, *Necator americanus* and *Ancylostoma duodenale*, and the whipworm, *Trichuris trichiura* (Chan *et al.* 1994; Bundy, 1997). In addition, more than 200 million people are estimated to harbour schistosome infections (WHO, 1993).

It is an epidemiological phenomenon that the distribution of helminths amongst hosts is over-dispersed. The majority of the hosts harbour few or no worms, whilst a minority of the hosts harbour much larger numbers of parasites (Anderson & Medley, 1985). This fact has clinical consequences for the host, as it is the intensity of infection that is the central determinant of the severity of morbidity (Stephenson, 1987; Cooper & Bundy, 1988). As

exemplified in Fig. 1, the age-intensity profile for *Trichuris trichiura* and *Ascaris lumbricoides* is typically convex with maximum intensity at 5–10 years of age. After peak intensity has been attained there is a dramatic decline in intensity to a low level, which then persists throughout adulthood.

A similar profile is apparent for *Schistosoma* infections, but with maximum intensity attained at a slightly later age, usually 10–14 years. A different profile is apparent for hookworm infections, as maximum intensity is usually not attained until 20–25 years (Stephenson, 1987). It is strikingly apparent that it is the school-age child who is particularly at risk from the clinical manifestations of disease. Indeed, it was estimated that, for girls and boys aged 5–14 years in low income countries, intestinal worms account for 12% and 11% of the total disease burden of this age group. An estimated 20% of disability adjusted life years (DALYs) lost due to communicable disease among school children are a direct result of intestinal nematodes (World Bank, 1993).

SO WHAT ARE THE CONSEQUENCES OF THESE INFECTIONS FOR CHILDREN?

Many children in low income countries underachieve and never realise their full potential. The aetiology of

* Corresponding author: Lesley Drake.
Tel: (44) 1865 281231. Fax: (44) 1865 281245.
E-mail: lesley.drake@ceid.ox.ac.uk

Parasitology (2001), **122**, S73–S81. Printed in the United Kingdom © 2001 Cambridge University Press

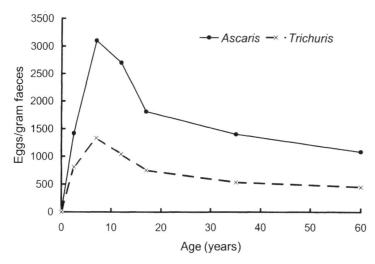

Fig. 1. Age intensity profile of *Ascaris lumbricoides* and *Trichuris trichiura*.

this underachievement is complex since numerous factors experienced during a child's formative years may have a detrimental effect upon development and functional efficiency. However there has been growing recognition of the specific deleterious effects of helminth infections upon both physical and intellectual development. As children suffer most at an age when they are both growing and learning, the entire developmental process is placed in jeopardy.

By far the most common effect on health is a subtle and insidious constraint on normal physical development, resulting in children failing to achieve their genetic potential for growth and suffering from the clinical consequences of iron deficiency anaemia and other nutritional deficiencies: heavy hookworm burdens have long been recognized as an important cause of iron deficiency anaemia (Roche & Layrisse, 1966); intense whipworm infection in children may result in *Trichuris* Dysentery Syndrome, the classical signs of which include growth retardation and anaemia (Bundy & Cooper, 1989); heavy burdens of both roundworm and whipworm are associated with protein energy malnutrition (PEM) (Stephenson *et al.* 1993).

In terms of intellectual development, there is an increasing body of evidence that these infections can have a detrimental effect on cognition and educational achievement in children (Sternberg, Powell & McGrane, 1997). The mechanism by which mental processes are affected is speculative. However, evidence lends itself to suggest that the mechanism is an indirect one. The most plausible mediators are the common sequelae of infection–iron deficiency anaemia (IDA) and undernutrition.

IDA in infants and young children is associated with significantly lower scores on psychological tests (Lozoff, Jimenez & Wolf, 1991; Pollitt *et al.* 1993; Lozoff, 1998). Deficits of 0·5–1·5 standard deviation units in scores on infant development scales or IQ tests for children have been found quite consistently

across studies and age groups. Moreover, these effects of IDA during infancy are associated with lower developmental test scores at 5 years of age (Cooper & Bundy, 1988). Similar effects have been shown during school age (Lozoff, 1998), although the relationship is less clear for children with mild IDA (Lozoff, Brittenham & Wolf, 1987). Treatment studies have consistently found improvements in cognitive function following iron supplementation in infancy (Idjradinata & Pollitt, 1993) and in middle childhood (Pollitt, Hathirat & Kotchabhakdi, 1989; Seshadri & Gopaldas, 1989; Soemantri, 1989; Soewendo, Husaini & Pollitt, 1989). IDA also leads to long-term deficits in cognitive functioning (Grantham-McGregor, Walker & Chang, 2000).

Studies of the effect of PEM on cognitive ability are strongly suggestive but not conclusive. Severe PEM is associated with cognitive deficits, but the relationship is not unequivocally causal (Simeon & Grantham-McGregor, 1990). In fact, when comparison groups are controlled for height-for-age (an indicator of long-term undernutrition), the effect of severe bouts of PEM disappears indicating that sustained undernutrition is important when considering effects on cognitive function (Grantham-McGregor, Powell & Fletcher, 1989). This conclusion is supported by studies of mild to moderate PEM, which demonstrate that height-for-age but not weight-for-age (a measure of short-term variations in nutritional input) is associated with development and achievement levels in school-children (Chun, 1971; Powell & Grantham-McGregor, 1980; Jamison, 1986; Moock & Leslie, 1986; Johnston, Low & De Baessa, 1987; Agarwal *et al.* 1987; Florencio, 1988) and in younger children (Powell & Grantham-McGregor, 1985).

The results of intervention (nutritional supplementation) studies are again inconclusive but appear to indicate that mild-moderate undernutrition has concurrent detrimental effects on cognitive function

(Grantham-McGregor, 1987). Severe malnutrition may lead to long-term cognitive deficits (Grantham-McGregor *et al.* 2000).

Whether helminth infections directly cause cognitive deficits is a matter of debate. The vast majority of over 40 studies demonstrate an association between geohelminthic infection and impaired cognitive or educational abilities (Waite & Nelson, 1919; De Carneri, Garofano & Grassi, 1968; Kvalsvig, Cooppan & Connolly, 1991; Nokes *et al.* 1991, 1992; Boivin *et al.* 1993; Callender *et al.* 1994; Nokes & Bundy, 1994; Simeon, Callender & Wong, 1994; Levav *et al.* 1995; Sternberg *et al.* 1997; Sakti *et al.* 1999). Only a small number of these studies, however, are well designed.

Cross-sectional studies are often flawed due to the proliferation of confounding variables. Infection prevalence is related to age, sex, and socio-economic status (for example, Sakti *et al.* 1999), among other variables. Many apparent differences between infected and uninfected children, particularly differences in educational test performance, disappear when confounding factors are controlled (Berlinguer, Del Trono & Orecchia, 1964; De Carneri *et al.* 1968). So, it is impossible to conclude from associative studies that cognitive impairments have been caused by infection, rather than simply having co-occurred with it.

One positive result to emerge from well designed treatment studies is that found by Nokes *et al.* (1992) in Jamaica. Children with moderate to heavy infections of whipworm showed improvements in working memory (measured by reciting strings of digits forwards and in reverse order) and rapid retrieval of words from long-term memory (verbal fluency) following treatment, as compared with a randomly assigned placebo group. No such improvements were found in visual short-term memory, problem-solving ability, cognitive speed, arithmetic, or verbal comprehension. Simeon, Grantham-McGregor & Wong (1995) studied a similar population with a slightly lower level of infection of whipworm. They found no main effect of treatment on performance on any of the cognitive tests they used, but they did find that wasted children (those with a low weight-for-age) showed improvements in verbal fluency after treatment. Another study using cognitive tests (Watkins, Cruz & Pollitt, 1996 *a*), this time in Guatemala, demonstrated improvements in the Sternberg short-term memory scanning test and in simple-choice reaction time (cognitive speediness) following treatment, but only for those children heavily infected with roundworm. Those lightly infected with the same parasite actually showed poorer performance on these tests after treatment. A similar negative effect of treatment was found by Pollit, Perez-Escamilla & Wayne (1991) in Kenya. In this study, children treated for hookworm infection were slower than were a randomly-selected placebo

group on a reaction-time task 8 months after treatment. No baseline measures were taken in this study, however, so it is difficult to determine whether the randomization process succeed in producing groups of children with equal abilities before treatment.

Two of the studies using a randomized control design also looked at education-related variables. An improvement in spelling and better attendance was observed with Jamaican children treated for whipworm (Berlinguer *et al.* 1964). However, the effect was evident only for stunted children (those with low height-for-age) with heavy worm burdens. Watkins *et al.* (1996 *b*) found no effect of treatment of children with moderate to heavy levels of roundworm on either reading or vocabulary score or on school attendance.

Studies often show an effect in the most heavily infected and affected children, but with such mixed results it is difficult to reach any strong conclusions as to whether helminthiases directly contribute to impairment of cognitive functioning (Dickson, Awsathi & Williamson, 2000). However, it is apparent that a pathway, via infection sequelae, does exist for these infections to at least contribute to this impairment (for a fuller review refer to Drake, Jukes & Sternberg, 2000), and that iron deficiency and PEM secondary to infection may be an important constraint.

Another consideration that should be made with regards to the impact of these infections on development, is that psychological development as a whole covers a broader range of abilities, not only including the more commonly studied domains of cognitive and motor development, but also social and emotional (such as understanding of interpersonal relationships, the development of control over one's emotions) and temperament (such as how children react to new situations). Such domains of psychological development are only just beginning to be included in assessment of the impact of ill health with recent studies of the effect of nutrient deficiency on affect and behaviour (Lozoff *et al.* 1998) and on stress (Fernald & Grantham-McGregor, 1998).

In addition, chronically infected children may become lethargic and less inclined to exert themselves and, as a consequence, not fully benefit from both their domestic and school environments. Given that these years may be only opportunity that many children have for formal education, by not taking full advantage of this they may compromise their social functioning and future potential.

Thus, these infections may also have an insidious and less overt impact upon the developmental process: delays in the development or impairment of these abilities may leave a child lacking in the interpersonal and emotional skills vital to making positive life decisions. This is particularly worrying given the current spread of the HIV/AIDS pandemic, and the importance of effective life skills at an early age in avoiding infection.

THE CONSEQUENCES ARE WORSE THAN PREVIOUSLY REALIZED

A growing body of evidence suggests that the effects of infection have tended to be underestimated. One important new factor is that the clinical consequences of infection can manifest themselves at much lower worm burdens than previously thought. Intervention studies have shown that an infection with as few as 10 roundworms is associated with deficits in growth and physical fitness in school-age children and moderate whipworm infections can cause growth retardation and anaemia in children (Cooper *et al.* 1993; Stephenson *et al.* 1933; Thein-Hlaing, 1993). Studies have also demonstrated that even light hookworm infection can lead to anaemia. This is evident not only in the adult population, but also in pre-school and school-children in certain populations (Stolzfus *et al.* 1997, 1998; Olsen *et al.* 1998; Beasley, Tompkins & Hall, 1999; Brooker, Booth & Guyatt, 1999; Brooker *et al.* 1999).

Children, adolescents and pregnant women are particularly at risk from hookworm morbidity as the physiological demands for iron are high during these periods. Evidence suggests that anaemia during pregnancy may adversely influence intrauterine growth rate, prematurity and birth weight. Thus, there may be important maternally-programmed consequences for children (Partnership for Child Development, 1997). In addition, it has been suggested that the clinical consequences of hookworm infection occur at a lower threshold in individuals who have a diet low in bioavailable iron (Crompton & Whitehead, 1993; Tatala, Svanberg & Mduma, 1998). Sadly, this is all too common an occurrence in communities in low income countries.

The second new factor is that multiple species infections may significantly exacerbate morbidity. In some species and regions, people with multiple infections are more common than those with either no infection or a single infection. For example, Crompton & Tulley (1987) listed 47 different protozoan and helminth species found in association with *A. lumbricoides* infections, of which 24 were common. The presence of a variety of human parasites that occur in the same environment each with a high prevalence, increases the likelihood of significant rates of co-infection. Another study estimated that multiple infections including *Plasmodium falciparum*, *Plasmodium malariae*, *Schistosoma haematobium* and three common nematodes occurred in approximately 60% of a Kenyan population (Ashford, Craig & Oppenheimer, 1992). There have been numerous other studies surveying

human infections and for a full review of these refer to Petney & Andrews (1998).

Howard, Donnelly & Chan (2000) found that, in general, associations between geohelminth species were positive, which implies that more multiple species infections are observed than would be expected by chance. Predisposition has been demonstrated for all the major helminth infection (Schad & Anderson, 1985; Elkins, Haswell-Elkins & Anderson 1986; Haswell-Elkins, Elkins & Anderson 1987; Holland *et al.* 1989; Bundy *et al.* 1987; Bundy & Cooper, 1988; Bradley & Chandiwana, 1990) and it has been suggested that perhaps certain individuals may be predisposed to multiple infections (Haswell-Elkins *et al.* 1987; Howard *et al.* 2000).

Behavioural factors could be responsible for multiple species infection. For example, *T. trichiura* and *A. lumbricoides* are both transmitted faeco-orally, thus behavioural factors that lead to exposure to and infection with one parasite, may also lead to exposure to and infection with the other parasite (Booth & Bundy, 1995). It has been noted that *A. lumbricoides* and *T. trichiura* infections can cluster within households (Forrester *et al.* 1988, 1990; Chan, Bundy & Khan, 1994) suggesting that individuals in the same peridomicillary environment share some risk factors for each infection. Whether this is caused by genetic factors, behavioural factors or a combination of the two has not been conclusively determined. Analyses do suggest that if there is a genetic basis for predisposition, its effects tend to be overwhelmed by other household-specific effects (Chan *et al.* 1994). However, it is now clear from segregation analysis that genetic factors do partly regulate susceptibility to *S. mansoni* infection (Marquet *et al.* 1996).

But how may multiple infections exacerbate morbidity? Studies suggest that certain morbid effects, for example, reduced cognitive function and reduced haemoglobin levels, may be exacerbated when individuals harbour certain combinations of infections (Robertson, 1992; Behnke *et al.* 1994). Observations indicate that individuals harbouring multiple species infections have higher than expected levels of each infection individually, and thus will be at a higher risk of morbidity (Booth *et al.* 1998). Needham, Kim & Cong (1998) demonstrated that high intensity *A. lumbricoides* infections were significantly more likely to occur with high intensity *T. trichiura* infections than would be expected by chance; children with multiple species infections have higher egg counts of each species than children with single species infections (Brooker, Miguel & Moulin, 2000).

Can we do anything to help these children?

The concept of health is broad. Health can be viewed in terms of the life cycle starting with pregnancy, moving through birth, infancy, childhood, the school years, adolescence, adulthood and ageing. Thus development and the benefit of health interventions are cumulative in effect – the benefit in one age group is at least partially dependent on interventions in an earlier age group. Prioritizing interventions at several points across the life cycle is necessary for sustained, high-impact improvements in health outcomes.

The major programmatic constraint in low income countries, where capacity and infrastructure are often lacking, is how to reach people in need. Success in reaching pregnant mothers and infants (under the age of five) has been attained through mother and child programmes and community based IMCI (integrated management of childhood illness) and ECD (early child development) initiatives. Children above the age of five can be most easily reached through school-based health and nutrition programmes, exploiting the existing educational infrastructure. By continuing interventions through the childhood years, the success of ECD and other early interventions can be built upon and strengthened.

Ensuring that children are healthy and able to learn is now recognized as an essential component of an effective education system. This is especially relevant to efforts to achieve 'Education For All' in the most deprived areas, as now more of the poorest and most disadvantaged children have access to school, many of whom are girls. It is these children, who are often the least healthy and most malnourished, who have the most to gain educationally from improved health.

Good health and nutrition are not only essential inputs but also important outcomes of basic education of good quality. On the one hand, children must be healthy and well-nourished in order to fully participate in education and gain its maximum benefits. Thus, programmes which improve health and nutrition can enhance the learning and educational outcomes of school children. On the other hand, quality education, including education about health, can lead to better health and nutrition outcomes for children and, especially through the education of girls, for the next generation of children as well.

Positive experiences by WHO, UNICEF, UNESCO and the World Bank suggest that there is a basic framework that could form the basis for an effective school health and nutrition programme upon which to build. This partnership seeks to Focus Resources on Effective School Health – *FRESH* Start. The framework includes four basic interventions: (1) school-based health policies; (2) provision of safe water and adequate sanitation; (3) skills based health education and (4) school-based health and nutrition services. These interventions are supported by an interactive framework of partnerships.

The cost-effectiveness of mass deworming as part of an effective school health and nutrition programme can be maximized by combining control of both schistosome and geohelminth infections, using two drugs: praziquantel and a benzimidazole derivative. Thus, multiple infections have important implications for the practical design of such programmes. Key to the design of such a strategy is the quantification of the pattern of multiple infections, as this approach will only be cost-effective when these infections are endemic in the same communities.

Preliminary analyses (Bundy *et al.* 1991) demonstrated that the geographical distributions of geohelminth and schistosome infections overlap sufficiently at the national level to warrant consideration for combined control. But it is also known that these infections are unevenly distributed within countries. Thus, methods to identify communities warranting a combined control approach, will enhance the efficacy of this approach.

Geographical information systems (GIS) are being used to develop predictive maps of infection prevalence rates. Focus is, at present, upon an infection atlas of Africa, but expansion is intended. Also, using overall estimates of species associations in different communities and countries, statistical predictive methods are being developed (Howard *et al.* 2000). This methodology is particularly useful when only marginal prevalence data are available.

A rapid assessment tool has been developed that can be used to identify target communities/schools warranting treatment of *Schistosoma haematobium* infections (Lengeler, Sala-Diakanda & Tanner, 1992). This is a questionnaire-based approach using self reporting of blood in urine as a reliable indicator of infection. It may also be possible to identify target communities for treatment of *S. mansoni* infections using a similar approach (Hailu, Jemaneh & Kebede, 1995; Booth, Mayombana & Kilima, 1998). A similar tool does not exist for detection of geohelminth infections. Brooker *et al.* (1999), attempted to predict the prevalence of geohelminth infection using schistosome infection prevalence. This approach proved unsuccessful as results suggested that schistosome and geohelminth infections are independently distributed within countries where they are both endemic. Reported gastro-intestinal symptoms also appear not to be associated with geohelminth infections (Booth *et al.* 1998). Thus, long-term cost-effectiveness of targetted combined control programmes may only benefit from the added expense of parasitological screening.

The combination of impact on physical growth and education gives helminth infection an importance to child development that far outweighs the apparent consequences of infection. A single, oral treatment is sufficient to significantly reduce worm burdens in all but the most intensely infected

populations. Mass delivery of anthelmintics is one of
the most cost-effective, simple and safe school-based
health services that can be delivered by trained
teachers. Evaluation of large-scale demonstration
school health programmes has shown that school-
based health services can have an impact on a broad
range of health and education outcomes (Partnership
for Child Development, 1998). For these reasons
deworming has been an essential component of
school health programmes in developing countries
(Savioli, Bundy & Tomkins, 1992).

REFERENCES

AGARWAL, D. K., UPADHYAY S. K., TRIPATHI A. M. &
  AGARWAL, K. N. (1987). Nutritional status, physical
  work capacity and mental function in school children.
  *Nutrition Foundation of India Scientific Report No. 6,
  New Delhi.* Nutrition Foundation of India.

ANDERSON, R. M. & MEDLEY, G. F. (1985). Community
  control of helminth infections of man by mass and
  selective chemotherapy. *Parasitology* **90**, 629–660.

ASHFORD, R. W., CRAIG, P. S. & OPPENHEIMER, S. J. (1992).
  Polyparasitism on the Kenya coast. 1. Prevalence, and
  association between parasitic infections. *Annals of
  Tropical Medicine and Parasitology* **86**, 671–679.

BEASLEY, N. M. R., TOMPKINS, A. M. & HALL, A.. (1999).
  The impact of population level deworming on
  haemoglobin levels in Tanga, Tanzania. *Tropical
  Medicine and International Health* **4**, 744–750.

BEHNKE, J. M., PRITCHARD, D. I., WAKELIN, D., PARK, J. R.,
  MCNICHOLAS, A. M. & GILBERT, F. S. (1994). Effect of
  ivermectin on infection with gastro-intestinal
  nematodes in Sierra Leone. *Journal of Helminthology*
  68, 187–195.

BERLINGEUR, G., DEL TRONO, L. & ORECCHIA, P. (1964).
  Eliminitasi intestinali, accrescimento scolastico (1964).
  Indagine sugli alumni della scuola D. Chiesa di Roma
  [Intestinal helminthiases, growth, and school
  achievement: A study among students of the
  D. Chiesa School in Rome]. *Rivista di Parassitologia*
  25, 77–92.

BOIVIN, M. J., GIORDANI, B., NDANGA, K., MAKY, M. M.,
  MANZEKI, K. M., NGUNU, N. & MUAMBA, K. (1993).
  Effects of treatment for intestinal parasites and
  malaria on cognitive abilities of school children in
  Zaire. *Health Psychology* **12**, 220–226.

BOOTH, M. & BUNDY, D. A. P. (1995). Estimating the
  numbers of multiple-species geohelminth infections in
  human communities. *Parasitology* **111**, 645–653.

BOOTH, M., BUNDY, D. A. P., ALBANICO, M., CHWAYA,
  H. M., ALAWI, K. S. & SAVIOLI, L. (1998). Associations
  among multiple geohelminth infections in
  schoolchildren from Pemba Island. *Parasitology* **116**,
  85–93.

BOOTH, M., MAYOMBANA, C. & KILIMA, P. (1998). The
  population biology and epidemiology of schistosome
  infections among schoolchildren in Tanzania.
  *Transactions of the Royal Society of Tropical Medicine
  and Hygiene* **92**, 491–495.

BRADLEY, M. & CHANDIWANA, S. K. (1990). Age-
  dependency in predisposition to hookworm infection

in the Burma valley area of Zimbabwe. *Transactions of
  the Royal Society of Tropical Medicine and Hygiene* **84**,
  826–828.

BROOKER, S., BOOTH, M. & GUYATT, H. L. (1999).
  Comparison of schistosome and geohelminth infection
  prevalences in school-aged children from selected
  areas of Africa: implications for rapid assessment and
  combined control. *Transactions of the Royal Society of
  Tropical Medicine and Hygiene* **93**, 125–126.

BROOKER, S., MIGUEL, E. & MOULIN, S. (2000).
  Epidemiology of single and multiple species of
  helminth infections among schoolchildren in Busia
  District, Kenya. *East African Medical Journal* **77**,
  157–161.

BROOKER, S., PESHU, N., WARN, P. A., MOSOBO, M.,
  GUYATT, H. L., MARSH, K. & SNOW, R. W. (1999). The
  epidemiology of hookworm infection and its
  contribution to anaemia among pre-school children on
  the Kenyan Coast. *Transactions of the Royal Society of
  Tropical Medicine and Hygiene* **93**, 240–246.

BUNDY, D. A. P. (1997). This wormy world – Then and
  now. *Parasitology Today* **13**, 407–408.

BUNDY, D. A. P., CHANDIWANA, S., HOMEIDA, M. M., YOON,
  S. & MOTT, K. E. (1991). The epidemiological
  implications of a multiple species approach to the
  control of human helminth infections. *Transactions of
  the Royal Society of Tropical Medicine and Hygiene* **85**,
  274–276.

BUNDY, D. A. P. & COOPER, E. S. (1988). The evidence for
  predisposition to trichuriasis in humans: comparison
  of institutional and community studies. *Annals of
  Tropical Medicine and Parasitology* **82**, 251–256.

BUNDY, D. A. P. & COOPER, E. S. (1989). *Trichuris* and
  trichuriasis in humans. *Advances in Parasitology* **28**,
  107–173.

BUNDY, D. A. P., COOPER, E. S., THOMPSON, D. E., DIDIER,
  J. M., ANDERSON, R. M. & SIMMONS, I. (1987).
  Predisposition to *Trichuris trichiura* infections in
  humans. *Epidemiology and Infection* **98**, 65–71.

CALLENDER, J. E., GRANTHAM-MCGREGOR, S. M., WALKER,
  S. P. & COOPER, E. S. (1994). Treatment effects in
  Trichuris dysentery syndrome. *Acta Paediatrica* **83**,
  1182–1187.

CHAN, L., BUNDY, D. A. P. & KAN, S. P. (1994).
  Aggregation and predisposition to *Ascaris lumbricoides*
  and *Trichuris trichiura* at the familial level.
  *Transactions of the Royal Society of Tropical Medicine
  and Hygiene* **88**, 46–8.

CHAN, M. S., MEDLEY, G. F., JAMISON, D. & BUNDY, D. A. P.
  (1994). The evaluation of potential global morbidity
  due to intestinal nematode infections. *Parasitology*
  109, 373–387.

CHUN, F. (1971). Nutrition and education – a study.
  Journal of Singapore Pediatric Society **13**, 91–96.

COOPER, E. S. & BUNDY, D. A. P. (1988). *Trichuris* is not
  trivial. *Parasitology Today* **4**, 301–306.

COOPER, E. S., DUFF, E. M. W., HOWELL, S. & BUNDY,
  D. A. P. (1995). "Catch up" growth velocities after
  treatment for *Trichuris* dysentery syndrome.
  *Transactions of the Royal Society of Tropical Medicine
  and Hygiene* **89**, 653.

CROMPTON, D. W. T. & TULLEY, J. J. (1987). How much
  ascariasis is there in Africa? *Parasitology Today* **3**,
  123–127.

CROMPTON, D. W. T. & WHITEHEAD, R. (1993). Hookworm infections and human iron metabolism. *Parasitology* **107**, S137–S145.

DE CARNERI, I., GAROFANO, M. & GRASSI., L. (1968). Il ruolo della tricocepalosi nel ritardo dello sviluppo frisco e mentale infantile in base ad un'inchiesta quantitative nel Basso Lodigiano [The role of trichuriasis in the physical and mental retardation of children based on a quantitative study in Basso Lodigiano]. *Rivista de Parassitologia* **28**, 103–122.

DICKSON, R., AWSATHI, S. & WILLIAMSON, P. (2000). Effects of treatment for intestinal helminth infection on growth and cognitive performance in children: systematic review of randomised trials. *British Medical Journal* **320**, 1697–1701.

DRAKE, L. J., JUKES, M. & STERNBERG, R. (2000). Geohelminthiases (ascariasis, trichuriasis and hookworm): cognitive and developmental impact. *Journal of Paediatric Infectious Disease* (in press).

ELKINS, D. B., HASWELL-ELKINS, M. & ANDERSON, R. M. (1986). The epidemiology and control of intestinal helminths in the Pulicat Lake region of Southern India. I. Study design and pre and post-treatment observations on *Ascaris lumbricoides* infection. *Transactions of the Royal Society of Tropical Medicine and Hygiene* **80**, 774–792.

FORRESTER, J. E., SCOTT, M. E., BUNDY, D. A. P. & GOLDEN, M. H. (1988). Clustering of *Ascaris lumbricoides* and *Trichuris trichuria* infections within households. *Transactions of the Royal Society of Tropical Medicine and Hygiene* **82**, 282–288.

FORRESTER, J. E., SCOTT, M. E., BUNDY, D. A. P. & GOLDEN, M. H. (1990). Predisposition of individuals and families in Mexico to heavy infection with *Ascaris lumbricoides* and *Trichuris trichiura*. *Transactions of the Royal Society of Tropical Medicine and Hygiene* **84**, 272–276.

FERNALD, L. C. & GRANTHAM-MCGREGOR, S. M. (1998). Stress response in school-age children who have been growth retarded since early childhood. *American Journal of Clinical Nutrition* **68**, 691–698.

FLORENCIO, C. (1988). *Nutrition, Health and Other Determinants of Academic Achievement and School-related Behaviour of Grades One to Six Pupils.* Quezan City, Philippines. University of the Philippines.

GRANTHAM-MCGREGOR, S. M. (1987). Field studies in early nutrition and later achievement. In *Early Nutrition and Later Achievement* (ed. Dobbing, J.), pp. 128–174. London: Academic Press.

GRANTHAM-MCGREGOR, S. M., POWELL, C. & FLETCHER, P. (1989). Stunting, severe malnutrition and mental development in young children. *European Journal of Clinical Nutrition* **43**, 403–409.

GRANTHAM-MCGREGOR, S. M., WALKER, S. P. & CHANG, S. (2000). Nutritional deficiencies and later behavioural development. *Proceedings of the Nutritional Society* **59**, 1–8.

HAILU, M., JEMANEH, L. & KEBEDE, D. (1995). The use of questionnaires for the identification of communities at risk for intestinal schistosomiasis in western Gojam. *Ethiopian Medical Journal* **33**, 103–113.

HASWELL-ELKINS, M. R., ELKINS, D. B. & ANDERSON, R. M. (1987). Evidence for predisposition in humans to infections with *Ascaris*, hookworm, *Enterobius* and *Trichuris* in a South Indian fishing community. *Parasitology* **95**, 323–327.

HOLLAND, C. V., ASAOLU, S. O., CROMPTON, D. W. T., STODDART, R. C., MACDONALD, R. & TORIMIRO, S. E. (1989). The epidemiology of *Ascaris lumbricoides* and other soil-transmitted helminths in primary school children from Ile-Ife, Nigeria. *Parasitology* **99**, 275–285.

HOWARD, S. C., DONNELLY, C. & CHAN, M. S. (2000). Methods for estimation of associations between multiple species parasite infections. *Parasitology* (in press).

IDJRADINATA, P. & POLLITT, E. (1993). Reversal of developmental delays among iron deficient anemic infants treated with iron. *Lancet* **341**, 1–4.

JAMISON, D. (1986). Child malnutrition and school performance in China. *Journal of Development Economics* **20**, 299–309.

JOHNSTON, F., LOW, S. & DE BAESSA, Y. (1987). Interaction of nutrition and socio-economic status as determinants of cognitive development in disadvantaged urban Guatemalan children. *American Journal of Physical Anthropology* **73**, 501–506.

KVALSVIG, J. D., COOPPAN, R. M. & CONNOLLY, K. J. (1991). The effects of parasite infections on cognitive processes in children. *Annals of Tropical Medicine and Parasitology* **85**, 551–568.

LENGELER, C., SALA-DIAKANDA, D. & TANNER, M. (1992). Using questionnaires through an existing administrative system: a new approach to health interview surveys. *International Journal of Epidemiology* **20**, 796–807.

LEVAV, M, MIRSKY, A. F., SCHANTZ, P. M., CASTRO, S & CRUZ, M. E. (1995). Parasitic infection in malnourished school children. *Parasitology* **110**, 103–111.

LOZOFF, B. (1998). Exploratory mechanisms for poorer development in iron-deficient anemic infants. In *Nutrition, Health and Child Development: Research Advances and Child Development* (ed. Grantham-McGregor, S. M.), pp. 162–178. Washington, DC: PAHO.

LOZOFF, B., BRITTENHAM, G. M. & WOLF, A. W. (1987). Iron deficiency anemia and iron therapy effects on infant developmental test performance. *Paediatrics* **79**, 981–995.

LOZOFF, B., JIMENEZ, E. & WOLF, A. W. (1991). Long-term developmental outcome of infants with iron deficiency. *New England Journal of Medicine* **325**, 687–694.

LOZOFF, B., KLEIN, N. K., NELSON, E. C., MCCLISH, D. K., MANUEL, M. & CHACON, M. E.. (1998). Behaviour of infants with iron deficiency anaemia. *Child Development* **1**, 24–36.

MARQUET, S., ABEL, L., HILLAIRE, D., DESSEIN, H., KALIL, J., FEINGOLD, J., WEISSENBACH, J & DESSEIN, A. J. (1996). Genetic localisation of a locus controlling the intensity of infection by *Schistosoma mansoni* on chromosome 5q31-q33. *Nature Genetics* **14**, 181–184.

MOOCK, P. & LESLIE, J. (1986). Childhood malnutrition and schooling in the Terai region of Nepal. *Journal of Development Economics* **20**, 33–52.

NEEDHAM, C. S., KIM, H. & CONG, L. (1998). Epidemiology of soil-transmitted nematodes in Ha Nam province,

Viet Nam. *Tropical Medicine and International Health* **3**, 904–912.

NOKES, C. & BUNDY, D. A. P. (1994). Does helminth infection affect mental processing and educational achievement? *Parasitology Today* **10**, 14–18.

NOKES, C., COOPER, E. S., ROBINSON, B. A. & BUNDY, D. A. P. (1991). Geohelminth infection and academic assessment in Jamaican children. *Transactions of the Royal Society of Tropical Medicine and Hygiene* **85**, 272–273.

NOKES, C., GRANTHAM-MCGREGOR, S. M., SAWYER, A. W., COOPER, E. S., ROBINSON, B. A. & BUNDY, D. A. P. (1992). Moderate to high infections of *Trichuris trichiura* and cognitive function in Jamaican school children. *Parasitology* **104**, 539–547.

OLSEN, A., MAGNUSSEN, P., OUMA, J. H., ANDREASSEN, J. & FRIIS, H. (1998). The contribution of hookworm and other parasitic infections to haemoglobin and iron status among children and adults in Western Kenya. *Transactions of the Royal Society of Tropical Medicine and Hygiene* **92**, 643–649.

PARTNERSHIP FOR CHILD DEVELOPMENT (1997). Better health, nutrition and education for the school-age child. *Transactions of the Royal Society of Tropical Medicine and Hygiene* **91**, 1–2.

PARTNERSHIP FOR CHILD DEVELOPMENT (1998). The health and nutritional status of schoolchildren in Africa: evidence from school-based programmes in Ghana and Tanzania. *Transactions of the Royal Society of Tropical Medicine and Hygiene* **92**, 254–261

PETNEY, T. N. & ANDREWS, R. H. (1998). Multiparasite communities in animals and humans: frequency, structure and pathogenic significance. *International Journal for Parasitology* **28**, 377–393.

POLLITT, E., GORMAN, K. S., ENGLE, P. L., MARTORELL, R. & RIVERA, J. (1993). Early supplementary feeding and cognition: effects over two decades. *Monographs of the Society for Child Development* **58**, 1–99.

POLLITT, E., HATHIRAT, P. & KOTCHABHAKDI, N. (1989). Iron deficiency and educational achievement in Thailand. *American Journal of Clinical Nutrition* **50** (supplement 3), 687–697.

POLLITT, E., PEREZ-ESCAMILLA, R. & WAYNE, W. (1991). Effects of infection with *Trichuris trichiura*, *Ascaris lumbricoides* and hookworm on information processing among Kenyan school children. *FASEB Journal* **5**, A1081.

POWELL, C. & GRANTHAM-MCGREGOR, S. M. (1980). The associations between nutritional status, school achievement and school attendance in twelve-year-old children at a Jamaican school. *West Indian Medical Journal* **29**, 247–253.

POWELL, C. & GRANTHAM-MCGREGOR, S. M. (1985). The ecology of nutritional status and development in young children in Kingston, Jamaica. *American Journal of Clinical Nutrition* **41**, 1322–1331.

ROBERTSON, L. J. (1992). Haemoglobin concentrations and concomitant infections with hookworm and *Trichuris trichiura* in Panamanian primary schoolchildren. *Transactions of the Royal Society for Tropical Medicine and Hygiene* **86**, 654–656.

ROCHE, M. & LAYRISSE, M. (1966). The nature and cause of hookworm anaemia. *American Journal of Tropical Medicine and Hygiene* **15**, 1029–1102.

SAKTI, H., NOKES, C., HERTANTO, W. S., HENDRATNO, S., HALL, A., BUNDY, D. A. P. & SATOTO. (1999). Evidence for an association between hookworm infection and cognitive function in Indonesian school children. *Tropical Medicine and International Health*. **4**, 322–334

SAVIOLI, L., BUNDY, D. & TOMKINS, A. (1992). Intestinal parasitic infections: a soluble public health problem. *Transactions of the Royal Society of Tropical Medicine and Hygiene* **86**, 353–354.

SCHAD, G. A. & ANDERSON, R. M. (1985). Predisposition to hookworm infection in humans. *Science* **228**, 1537–1540.

SESHADRI, S. & GOPALDAS, T. (1989). Impact of iron supplementation on cognitive function in pre-school and school-age children: the Indian experience. *American Journal of Clinical Nutrition* **50**, 675–686.

SIMEON, D. T., CALLENDER, J. & WONG, M. S. (1994). School performance, nutritional status and trichuriasis in Jamaican schoolchildren. *Acta Paediatrica* **83**, 1188–1193.

SIMEON, D. T. & GRANTHAM-MCGREGOR, S. M. (1990). Nutritional deficiency and children's behaviour and mental development. *Nutritional Research Review* **3**, 1–24.

SIMEON, D. T., GRANTHAM-MCGREGOR, S. M. & WONG, M. S. (1995). *Trichuris trichiura* infection and cognition in children: results of a randomised clinical trial. *Parasitology* **110**, 457–464.

SOEMANTRI, A. G. (1989). Preliminary findings on iron supplementation and learning achievement of rural Indonesian children. *American Journal of Clinical Nutrition* **50**, 698–702.

SOEWONDO, S., HUSAINI, M. & POLLITT, E. (1989). Effects of iron deficiency on attention and learning processes in preschool children: Bandung, Indonesia. *American Journal of Clinical Nutrition* **50**, 667–674.

STEPHENSON, L. S. (1987). *Impact of Helminth Infections on Human Nutrition: Schistosomes and Soil Transmitted Helminths*. London: Taylor & Francis.

STEPHENSON, L. S., LATHAM, M., ADAMS, E. J., KINOTI, S. N. & PERTET, A. (1993). Physical fitness, growth and appetite of Kenyan schoolboys with hookworm, *Trichuris trichiura* and *Ascaris lumbricoides* infections are improved four months after a single dose of albendazole. *Journal of Nutrition* **123**, 1036–1046.

STERNBERG, R. J., POWELL, C. & MCGRANE, P. (1997). Effects of a parasitic infection on cognitive functioning. *Journal of Experimental Psychology* **3**, 67–76.

STOLTZFUS, R. J., ALBONICO, M., CHWAYA, H. M., TIELSCH, J. M., SCHULZE, K. J. & SAVIOLI, L. (1997). Epidemiology of iron deficiency anaemia in Zanzibari schoolchildren: the importance of hookworms. *American Journal of Clinical Nutrition* **65**, 153–159.

STOLTZFUS, R. J., ALBONICO, M., CHWAYA, H. M., TIELSCH, J. M., SCHULZE, K. J. & SAVIOLI, L. (1998). The effects of the Zanzibar school-based deworming programme on iron status of children. *American Journal of Clinical Nutrition* **68**, 179–186.

TATALA, S., SVANBERG, U. & MDUMA, B. (1998). Low dietary iron is a major cause of anaemia: a nutrition survey in the Lindi District of Tanzania. *American Journal of Clinical Nutrition* **68**, 171–178.

THEIN-HLAING (1993). Ascariasis and childhood nutrition. *Parasitology* **107**, S125–S136.

WAITE, J. H. & NELSON, I. L. (1919). Study of the effects of hookworm infection upon the mental development of North Queensland schoolchildren. *Medical Journal of Australia* **1**, 1–8.

WATKINS, W. E., CRUZ, J. R. & POLLITT, E. (1996a). The effects of deworming on indicators of school performance in Guatemala. *Transactions of the Royal Society of Tropical Medicine and Hygiene* **90**, 156–162.

WATKINS, W. E., CRUZ, J. R. & POLLITT, E. (1999b). Whether deworming improves or impairs information

processing depends on intensity of *Ascaris* infection. Yale-China Association, Shatin, New Territories, Hong Kong.

WARREN, K. S., BUNDY, D. A. P. & ANDERSON, R. M. (1993). Helminth Infection. In *Disease Control Priorities in Developing Countries*. (ed. Jamison, D. T.), pp. 131–16. Oxford: Oxford University Press.

WHO (1993). The control of schistosomiasis. Second report of the WHO Expert Committee. *Technical Report Series* **830**, Geneva: WHO.

WORLD BANK (1993). World Development Report: Investing in Health. Oxford: Oxford University Press.

Chemotherapy for patients with multiple parasitic infections

P. L. CHIODINI*

Department of Clinical Parasitology, The Hospital for Tropical Diseases, London WC1E 6AU

SUMMARY

Multiple parasitic infections have become increasingly recognized as a result of improvements in laboratory diagnosis and a growing population of immunocompromised individuals. This review examines the principles of chemotherapy in groupings of multiple infections which are of particular clinical significance.

Key words: Multiple parasites, *Onchocerca*, *Loa loa*, intestinal helminths, schistosomiasis, malaria, AIDS, microsporidia, chemotherapy.

INTRODUCTION

Conventional medical training teaches medical students to make a single unifying diagnosis when analysing a clinical problem. Parasitology is an important exception to this rule and, especially in the tropics, it is common for an individual patient to harbour more than one parasite, sometimes in different organ systems or, especially in the case of the gastrointestinal tract, in the same one (Kang *et al.* 1998).

The approach to chemotherapy of multiple parasitic infections depends upon the following principles. (1) Precise diagnosis. At times this can be surprisingly difficult – in tropical areas diagnostic laboratory provision can vary from nil to sophisticated, but is most denied to those most affected by parasitic disease. This has led to syndromic treatment protocols and empirical, often broad-spectrum, antiparasitic therapy. In affluent countries with well equipped laboratory services, lack of experience can still lead to difficulties in confirming the diagnosis outside reference centres, though external quality assessment schemes, e.g. UKNEQAS, have raised standards in recent years (Kettelhut *et al.* 1994). (2) The choice of narrow versus broad-spectrum agents is influenced by access to accurate laboratory diagnosis and by cost (both of diagnostic tests and the drugs required). (3) Whether an individual or the community is being treated. (4) The possibility of drug interaction if two or more antiparasitic agents are given simultaneously for multiple parasitic infections. (5) Where chemotherapy for one parasite might produce complications due to its action on another parasite present simultaneously. (6) In an individual case, which parasite predominates as the cause of disease.

* E-mail: peter.chiodini@uclh.org

Multiple parasitic infections can occur in many potential combinations but there are certain groupings which are of particular significance and are considered below.

ONCHOCERCIASIS AND LOIASIS

When ivermectin was under consideration as part of the Onchocerciasis Control Programme (OCP), its safety in community-based mass treatment programmes in geographical areas where the population harbours multiple filarial infections plus intestinal nematodes, was clearly critical to the acceptability and potential success of the mass treatment programme.

Richard-Lenoble *et al.* (1988) examined the safety and efficacy of ivermectin in patients with multiple filarial infections in Gabon. Seventeen patients with concomitant *Loa loa* and *Onchocerca volvulus* were studied. Other nematode infections found in this group were *Mansonella perstans* (five patients) and intestinal nematodes (16 patients). Each patient received ivermectin 200 μg/kg as a single dose. Ten days later, the mean *Loa loa* microfilarial count had fallen to twenty per cent of the mean level before treatment, and *Onchocerca volvulus* dermal microfilarial densities were only 2% of the pre-treatment values, whilst *Mansonella perstans* microfilarial counts were unaffected by ivermectin. By day 23, 15/15 with *Ascaris* infection had been parasitologically cured, but there had been no significant effect upon infections with *Trichuris trichiura* or *Necator americanus*. Ten (59%) patients suffered pruritis one to five days post treatment, though neither antihistamines nor corticosteroids were required for symptomatic relief. However, the mean *Loa loa* microfilarial count was less than 300/ml of blood, so the tolerance of ivermectin with high microfilarial counts could not be determined by this study.

In 1995, the African Programme for Onchocerciasis Control (APOC) was launched, to develop community-based ivermectin treatment programmes in the nineteen *Onchocerca volvulus* endemic African countries outside the Onchocerca Control Programme in West Africa. Of the 19 participating countries, 12 have foci of *Loa loa* infection. Gardon *et al.* (1997) studied the incidence of serious adverse events following country-based ivermectin treatment in a part of Cameroon where onchocerciasis and loaisis were both endemic and explored the relationship between serious adverse events and pre-treatment *Loa loa* microfilarial counts. Of 17877 patients treated, 20 developed serious non-neurological reactions, one of them fatal. Two patients developed serious neurological reactions, including coma, from which they had recovered after a month. Both had pre-treatment microfilarial counts in excess of 50000/ml. It was possible to record the pre-treatment microfilarial count for 5500 patients and the initial *Loa loa* microfilarial load was the main risk factor for the development of serious reactions; the risk increasing with microfilarial intensity. The association between microfilarial load and the occurrence of marked or serious reactions was significant above 8000 microfilaria/ml of blood. The occurrence of serious reactions was closely related to the *Loa loa* microfilarial load (odds ratio 114·7). The authors noted three previous cases of *Loa loa* encephalopathy related to ivermectin, also from Cameroon, and estimated the incidence of serious neurological reactions to be approximately 1·1 per 10000 people. The authors concluded that in areas where loiasis and onchocerciasis co-exist, the use of ivermectin should be carefully considered if the onchocerciasis does not present a serious public health problem locally. In other areas, where onchocerciasis is blinding or severely disabling, persons at risk need to be identified. Ivermectin appears able to provoke the passage of microfilariae of *Loa loa* into the cerebrospinal fluid (CSF) (Ducorps *et al.* 1995). A probable case of *Loa loa* encephalopathy related to ivermectin (PLERI) has been defined by four criteria: occurrence of a coma in a person who was previously healthy and has no other cause for the coma; onset of CNS symptoms and signs within 5 days of ivermectin therapy and progression to coma without remission; *Loa loa* microfilaraemia of greater than or equal to 10 000 microfilariae/ml pre-treatment or greater than or equal to 1000 microfilariae per ml within two months after treatment; the presence of *Loa loa* microfilariae in the CSF.

Boussinesq *et al.* (1998) gave detailed descriptions of the PLERI cases noted by Gardon *et al.* (1997) which occurred in an area of Cameroon where 40–95 % of the population had dermal microfilariae of *Onchocerca* and 10–33% had *Loa loa* micro-

filaraemia. Of 17 877 persons treated with ivermectin 150 μg/kg, two developed encephalopathy and twenty developed milder neurological reactions, with functional impairment requiring assistance to perform everyday domestic activities, but without disorders of consciousness or neurological signs. Boussinesq *et al.* (1998) reviewed a total of five PLERI cases. All occurred in young, previously healthy males. Initial symptoms, including fatigue, headache and joint pains, appeared on day one or two post-ivermectin. Disordered consciousness was usually manifest on day three or four. The patients were usually incontinent for several days, but the motor deficit was usually mild. Tendon reflexes were absent in three and brisk in one case; cog-wheeling was seen in two patients. All three individuals whose fundi were examined had retinal haemorrhages. Two patients died, one from gastrointestinal haemorrhage and one as a result of secondary bacterial infection. In both cases the diagnosis of PLERI was made too late for them to receive adequate therapy. In the three survivors, the clinical picture was most severe on days four or five, then showed progressive improvement, with almost full recovery one month after ivermectin therapy. Peripheral blood microfilarial counts were very high in all five PLERI cases and all had microfilariae in the CSF.

The major risk factor for PLERI is the peripheral blood microfilarial count (a level above 30000 microfilariae/ml has been quoted as the risk threshold, (Chippaux *et al.* 1996) though a variety of co-factors has been suggested.

Treatment of PLERI involves good supportive care in hospital, antihistamines and/or cortico-steroids, and early detection and treatment of secondary infection (Boussinesq *et al.* 1998). Prevention of PLERI depends upon identification of individuals at risk from a high microfilarial load; a difficult task without taking pre-treatment blood samples from the population in areas with a significant overlap of *Onchocerca volvulus* and *Loa loa* infections.

INTESTINAL HELMINTHS

In cases of mixed intestinal helminthic infections, the effect on the human host is influenced by the relative intensity of the infections present. For example, Sakti *et al.* (1999) studied cognitive functioning in a population of Indonesian school children in whom there was a 43% prevalence of helminth infection (24% hookworm; 24% *Trichuris trichiura*; 8% *Ascaris lumbricoides*). Lower cognitive test scores were most apparent in children infected with hookworm rather than other helminths. The authors suggested this may be because the prevalence and/or intensity of infection with *Ascaris* or *Trichuris* were sufficiently low so as not to be responsible for significant morbidity or a public health problem.

This contrasts with other studies where *Trichuris* infection has exerted a significant influence on cognitive test scores.

As hookworm, *Ascaris* and *Trichuris* all respond to single dose (400 mg) albendazole therapy (Johnson & Soave, 1999), in cases of mixed infection there is no clinical need to determine which of them is acting as the predominant pathogen either in an individual or community treatment setting. However, uptake of therapy is a major determinant of treatment success or failure, with a tendency for polypharmacy to result in reduced compliance. Thus, where the pattern of mixed infection dictates the use of two or more different drugs, e.g. praziquantel for schisto-somiasis and albendazole for hookworm, a common delivery system should be deployed (Lwambo *et al*. 1999).

Intestinal nematodes and protozoa often coincide in infections of the gastrointestinal tract and Penggabean *et al*. (1998) examined the efficacy of albendazole on *Trichuris* and *Giardia* infections in rural Malaysia. Albendazole 400 mg daily for three days was followed by cure rates of 91·5% for *T. trichiura* and 96·6% for *G. intestinalis*. Reynoldson *et al*. (1998) studied the effect of albendazole, 400 mg daily for 5 days, on giardiasis and hookworm infection in an Aboriginal community in Western Australia. The prevalence of *Giardia* fell from 36·6% pre-treatment to 12% between days 6 and 9, 15% for days 10 to 17 and rose to 28% between days 18 and 30. Hookworm prevalence fell from 76% pre-albendazole to 2% between days 6–9 and was zero by days 18–30.

Community anthelminthic therapy is reviewed in detail in the chapter by Drake and Bundy in this supplement to which the reader is referred. There is still much work to do in studying the effects of treating geohelminth infection (multiple or single) in children and there will be no shortage of debate. For example, a recent Cochrane review concluded that public health investments in routine treatment of children in areas where helminths are common, based on the expectation of improved growth and learning, are not based on consistent or reliable evidence (Dickson *et al*. 2000).

SCHISTOSOMIASIS

Recognition that co-infection with schistosomes and intestinal helminths was common in various parts of Africa (Adewunmi *et al*. 1993; Birrie, Erko & Tedla 1994) has been followed by studies to assess the safety and efficacy of simultaneous combined mass treatment programmes.

Nokes *et al*. (1999) studied the effects of combined albendazole (400 mg single dose) and praziquantel (60 mg/kg split dose 3 h apart) for the treatment of geohelminth infections and *Schistosoma japonicum* infections respectively. Neither drug affected the cure rate of the other drug and co-administration of these agents was both safe and effective. Beasley *et al*. (1999) demonstrated haematological benefits (reduced fall in haemoglobin) from single dose albendazole (400 mg) and praziquantel (40 mg/kg) treatment compared to placebo in almost half of school children co-infected with geohelminths and *Schistosoma haematobium* in Tanga, Tanzania.

Schistosomiasis has also been associated with relapses of enteric fever (Gendrel *et al*. 1994), and persistent *Salmonella* bacteraemia in AIDS (Lambertucci, Rayes & Gerspacher-Lara 1998); persistent *Salmonella* bacteraemia appears to ex-acerbate a pre-existing sub-clinical schistosomal glomerulopathy (Martinelli *et al*. 1992). Acute schistosomiasis has been reported to facilitate the development of *Staphylococcus aureus* liver abscesses (Lambertucci *et al*. 1998). Detailed discussion of these issues is outside the remit of this review, but effective treatment for schistosomiasis is required in each case.

MALARIA

Most infections with the human malaria parasites *Plasmodium falciparum*, *P. vivax*, *P. ovale* and *P. malariae* occur as single species. In cases where a patient harbours more than one species of malaria parasite, one or other species tends to be more numerous in the blood film, such that identification of the second species by microscopy can be very difficult, especially when young ring stages pre-dominate in the blood film under examination. The advent of molecular techniques for the detection of malaria parasites has led to greater appreciation of the importance of mixed malarial infections. Brown *et al*. (1992) conducted a prospective comparison of microscopic diagnosis and PCR for the circum-sporozoite gene for *Plasmodium falciparum* and for *P. vivax* in Thai soldiers. Of 137 consecutive cases of malaria, 3/32 (9%) of microscopically diagnosed *P. falciparum* infections and 5/104 (5%) of micro-scopically diagnosed *P. vivax* infections were found to be mixed by PCR. One case was diagnosed as mixed by microscopy and PCR. Pieroni *et al*. (1998) reported on 148 travellers with malaria as determined by PCR and microscopy, six (4·1%) of whom were shown to have mixed infections.

Treatment of mixed malarial infections is species dependent. In all cases the first priority is to treat the asexual erythrocytic stages as they are responsible for the malarial illness. If *P. falciparum* is present in the mixed infection it should be accorded priority as the most dangerous of the species infecting humans. The agents used are determined by the geographical origin of the parasite and thus its likely drug sensitivity/resistance profile, and the route of administration by the level of parasitaemia, severity of illness, the presence of complications and the ability to swallow oral medication in the patient concerned (WHO, 2000). Schizonticides effective against *P. falciparum* are usually effective against *P. vivax*, *P. ovale* and *P. malariae*. The asexual stages of mixed infections containing any combination of *P. vivax*, *P. ovale* or *P. malariae* can be treated with chloroquine (or quinine if the combination is thought to include chloroquine-resistant *P. vivax*). Mixed infections which include *P. vivax* or *P. ovale* require treatment directed against hypnozoites in the liver, to effect a radical cure and prevent relapse. The only agent currently marketed for this indication is the 8-aminoquinoline, primaquine (Anon, 1991). The standard dosage regimen is 7·5 mg twice daily for 14 days, but use of a higher dose (15 mg twice daily for 14 days) has been advocated following reports of treatment failure at the standard dose (Doherty *et al.* 1997). It is essential to check the glucose-6-phosphate-dehydrogenase (G6PD) status of the patient prior to administering primaquine to avoid causing haemolysis in G6PD-deficient individuals.

Malaria and helminths

Tshikuka *et al.* (1996) studied 1100 children and mothers in Lubumbashi, Zaire to survey the relationship between parasitic infections and clinical syndromes to see whether single and/or multiple species infections were risk factors for the clinical syndromes concerned. They found no significant interactions between *Plasmodium*, *Ascaris lumbricoides*, *Trichuris* and hookworm and no evidence of synergism or antagonism between the parasites present in relation to the resulting disease. In such a situation, management of the parasitic infections detected can proceed by treating the most dangerous parasite first, though simultaneous treatment would also be acceptable in the absence of a risk of drug interaction. Fryauff *et al.* (1998) examined the effect of chloroquine or primaquine anti-malarial prophylaxis given for one year for malaria prevention on the presence of intestinal parasitic infections in a group of Javanese men in Irian Jaya. They found no significant change between baseline and endpoint in the type of species found, the mean number of species or ova per subject, the relative proportion of infections caused by these species or the occurrence of parasite free, single or multiple infections.

Malaria and other protozoa

Malarone® (atovaquone/proguanil), a potent new antimalarial, also has activity against *Toxoplasma* and *Pneumocystis carinii* (via the atovaquone component in each case), so administration of this agent will have a wider antiparasitic effect than would be the case with more traditional antimalarials (such as quinine) which have a narrower spectrum of activity. Most agents used for antimalarial chemoprophylaxis

(with the exception of proguanil which has some causal prophylactic activity) act as schizonticides to clear low grade asexual parasitaemia. Malarone® has both schizonticide and causal prophylactic activity (against pre-erythrocytic stages of all four species of human malaria parasites) but still lacks activity against hypnozoites, a property possessed only by primaquine and tafenoquine of the agents available or on clinical trial at present. Thus, if a person is challenged by multiple species of malaria parasites which include *P. vivax* or *P. ovale*, a risk of a delayed primary attack of malaria remains, unless primaquine or tafenoquine is used as a prophylactic agent.

In areas of high malaria endemicity, where there is a large proportion of individuals semi-immune to malaria, many people are asymptomatic but parasitaemic. In such a situation it can be difficult to decide whether or not malarial parasitaemia is contributing to the presenting clinical syndrome (e.g. fever and cough) where a mixed infection may be present. Keeping an open mind and looking, or treating if appropriate, for other causes, whilst also treating the malaria parasitaemia, presents a suitably pragmatic approach to this problem.

An association between falciparum malaria and *Salmonella* bacteraemia has been noted by some authors (Gopinath, Keystone & Kain 1995; Ammah *et al.* 1999; Graham *et al.* 2000) but felt to be overestimated by others (Enwere *et al.* 1998). Full discussion of the issue is outside the scope of this review.

MULTIPLE PARASITIC INFECTIONS IN AIDS

The appearance of the Acquired Immune Deficiency Syndrome (AIDS) had a dramatic effect on parasitology. The previously obscure microsporidia were implicated in the causation of refractory diarrhoea in late stage AIDS (Hewan-Lowe *et al.* 1997; Kotler & Orenstein, 1998) and cerebral toxoplasmosis, isosporiasis and cryptosporidiosis all came to prominence in this condition. Pneumonia due to *Pneumocystis carinii* (now classified amongst the fungi) became an indicator for an AIDS diagnosis. Given such an environment of diminished immunity it is not surprising that a variety of parasites appeared as co-infections and that treatment, maintenance therapy or prophylaxis directed against one parasite had an effect on other organisms present. Fortunately, this can be turned to the patient's advantage. Co-trimoxazole used for prophylaxis against *Pneumocystis carinii* also has prophylactic activity against *Toxoplasma gondii* and is effective for the treatment and maintenance therapy of *Isospora belli* and *Cyclospora cayetanensis*. However, there may be concerns arising from the use of cross-reactive antiparasitic drugs. For example, atovaquone, a second line agent for the treatment of toxoplasmosis (Fung & Kirschenbaum, 1996; Katlama *et al.* 1996) or the prevention and treatment of pneumonia due to *Pneumocystis carinii* (Castro, 1998; El-Sadr *et al.* 1998), has good antimalarial activity, but *Plasmodium falciparum* infections readily recrudesce when treated with atovaquone alone (Chiodini *et al.* 1995), hence its use in combination with proguanil as the product Malarone®, when it is used against malaria. If atovaquone is deployed against *Toxoplasma* or *Pneumocystis* in individuals exposed to frequent malaria challenge, there is a potential to encourage the development of atovaquone-resistant *Plasmodium falciparum*.

Nowhere is the issue of unequal access to precise laboratory diagnosis and targetted therapy more evident than in the investigation of AIDS-associated diarrhoea. Where the techniques are available, more than 80% of diarrhoea cases in patients with HIV/AIDS can be attributed to a specific enteropathogen and *Cryptosporidium parvum*, *Isospora belli*, *Cyclospora cayetanensis* and the microsporidia explain at least 50% of cases of persistent diarrhoea (Farthing, Kelly & Veitch 1996).

Albendazole has an unusually wide spectrum for an antiparasitic drug, with clinically useful activity against larval cestodes (cysticercosis, hydatid disease), *Giardia*, some microsporidia (Molina *et al.* 1998), intestinal nematodes and lymphatic filariasis. This wide spectrum helps compliance with treatment in a condition in which multiple drug therapy is common. Its activity against the microsporidian *Enterocytozoon intestinalis*, *Giardia* and intestinal nematodes is particularly useful in the management of AIDS-associated diarrhoea in situations in situations where precise laboratory diagnosis may not be possible. For example, Kelly *et al.* (1996) examined the effect of a 2 week, high dose (800 mg twice daily) albendazole course versus placebo on persistent diarrhoea of more than three weeks' duration in 174 HIV seropositive patients in Zambia. No gastrointestinal investigations were performed. The patients who received albendazole had diarrhoea on 29% fewer days than the placebo group ($P < 0.0001$) in the two weeks after therapy and the therapeutic benefit of albendazole was maintained over six months, supporting its use as empirical therapy in the context described.

One of the most useful benefits from treating co-infections in AIDS has come not from an antiparasitic drug, but from antiviral chemotherapy directed against the underlying HIV infection. The introduction of combination highly active anti-retroviral therapy (HAART) has improved the ability of patients to control AIDS-associated opportunistic infections. For example, Conteas *et al.* (1998) looked at modification of the clinical course of intestinal microsporidiosis due to *Enterocytozoon bieneusi* in AIDS patients according to their immune status and anti-retroviral therapy. Decreased time to clearance of *E. bieneusi* was associated with peripheral blood CD4 cell counts greater than or equal to 100 mm^3, the use of two or more antiretroviral drugs, and use of a protease inhibitor.

CONCLUSIONS

There is likely to be increasing recognition of the presence and importance of multiple parasitic infections in human hosts as more sensitive, highly sophisticated laboratory techniques for their detection become more readily available and as the population of immunocompromised individuals increases.

Broad-spectrum agents, or single delivery systems for administration of multiple agents, will form the mainstay of mass treatment programmes for intestinal parasites conducted without prior laboratory diagnosis. Caution must always be exercised in planning large scale antiparasitic treatment programmes, whatever the parasite or organ system involved, to avoid the potential for adverse events related to drug activity on concomitant infections.

REFERENCES

ADEWUNMI, C. O., GEBREMEDHIN, G., BECKER, W., OLURUNMOLA, F. O., DORFLER, G. & ADENWUNMI, T. A. (1993). Schistosomiasis and intestinal parasites in rural villages in southwest Nigeria: an indication for expanded programme on drug distribution and integrated control programme in Nigeria. *Tropical Medicine and Parasitology* **44**, 177–180.

AMMAH, A., NKUO-AKENJI, T., NDIP, R. & DEAS, J. E. (1999). An update on concurrent malaria and typhoid fever in Cameroon. *Transactions of the Royal Society of Tropical Medicine and Hygiene* **93**, 127–129.

ANON (1991). Primaquine. In *Therapeutic Drugs* (ed. Dollery, C.) Edinburgh: Churchill Livingstone

BEASLEY, N. M. R., TOMKINS, A. M., HALL, A., KIHAMIA, C. M., LORRI, W., NDUMA, B., ISSAE, W., NOKES, C. & BUNDY, D. A. P. (1999). The impact of population level deworming on the haemoglobin levels of schoolchildren in Tanga, Tanzania. *Tropical Medicine and International Health* **4**, 744–750.

BIRRIE, H., ERKO, B. & TEDLA, S. (1994). Intestinal helminthic infections in the southern Rift Valley of Ethiopia with special reference to schistosomiasis. *East African Medical Journal* **71**, 447–452.

BOUSSINESQ, M., GARDON, J., GARDON-WENDEL, N., KAMGNO, J., NGOUMOU, P. & CHIPPAUX, J.-P. (1998). Three probable cases of *Loa loa* encephalopathy following ivermectin treatment for onchocerciasis. *American Journal of Tropical Medicine and Hygiene* **58**, 461–469.

BROWN, A. E., KAIN, K. C., PIPITHKUL, J. & WEBSTER, H. K. (1992). Demonstration by the polymerase chain reaction of mixed *P. falciparum* and *P. vivax* infections undetected by conventional microscopy. *Transactions of the Royal Society of Tropical Medicine and Hygiene* **86**, 609–612.

CASTRO, M. (1998). Treatment and prophylaxis of *Pneumocystis carinii* pneumonia. *Seminars in Respiratory Infection* **13**, 296–303.

CHIODINI, P. L., CONLON, C. P., HUTCHINSON, D. B., FARQUHAR, J. A., HALL, A. P., PETO, T. E. & BIRLEY, H. (1995). Evaluation of atovaquone in the treatment of patients with uncomplicated *Plasmodium falciparum* malaria. *Journal of Antimicrobial Chemotherapy* **36**, 1073–1078.

CHIPPAUX, J. P., BOUSSINESQ, M., GARDON, J., GARDON-WENDEL, N. & ERNAULD, J. C. (1996). Severe adverse reaction risks during mass treatment with ivermectin in loiasis-endemic areas. *Parasitology Today* **12**, 448–450.

CONTEAS, C. N., BERLIN, O. G. W., SPECK, C. E., PANDHUMAS, S. S., LARIVIERE, M. J. & FU, C. (1998). Modification of the clinical course of intestinal microsporidiosis in acquired immunodeficiency syndrome patients by immune status and anti-human immunodeficiency virus therapy. *American Journal of Tropical Medicine and Hygiene* **58**, 555–558.

DICKSON, R., AWASTHI, S., DEMELLWEEK, C. & WILLIAMSON, P. (2000). Anthelmintic drugs for treating worms in children: effects on growth and cognitive performance (Cochrane Review) In *The Cochrane Library* **Issue 4**. Oxford, Update Software

DOHERTY, J. F., DAY, J. H., WARHURST, D. C. & CHIODINI, P. L. (1997). Treatment of *Plasmodium vivax* malaria – time for a change? *Transactions of the Royal Society of Tropical Medicine and Hygiene* **91**, 76.

DUCORPS, M., GARDON-WENDEL, N., RANQUE, S., NDONG, W., BOUSSINESQ, M., GARDON, J., SCHNEIDER, D. & CHIPPAUX, J. P. (1995). Effets secondaires du traitement de la loase hypermicrofilaremique par l'ivermectine. *Bulletin de la Société de Pathologie Exotique* **88**, 105–112.

EL-SADR, W. M., MURPHY, R. L., YURIK, T. M., LUSKIN-HAWK, R., CHEUNG, T. W., BALFOUR, H. H., ENG, R., HOOTON, T. M., KERKERING, T. M., SCHUTZ, M., VAN DER HORST, C. & HAFNER, R. (1998). Atovaquone compared with dapsone for the prevention of *Pneumocystis carinii* pneumonia in patients with HIV infection who cannot tolerate trimethoprim, sulfonamides or both. Community Program for Clinical Research on AIDS and the AIDS Clinical Trials Group. *New England Journal of Medicine* **339**, 1889–1895.

ENWERE, G., VAN HENSBROEK, M. B., ADEGBOLA, R., PALMER, A., ONYIORA, E., WEBER, M. & GREENWOOD, B. (1998). Bacteraemia in cerebral malaria. *Annals of Tropical Paediatrics* **18**, 275–278.

FARTHING, M. J. G., KELLY, M. P. & VEITCH, A. M. (1996). Recently recognised microbial enteropathies and HIV infection. *Journal of Antimicrobial Chemotherapy* **37**, Suppl. B, 61–70.

FRYAUFF, D. J., PRODJODIPURO, P., BASRI, H., JONES, T. R., MOUZIN, E., WIDJAJA, H. & SUBIANTO, B. (1998). Intestinal parasite infections after extended use of chloroquine or primaquine for malaria prevention. *Journal of Parasitology* **84**, 626–629.

FUNG, H. B. & KIRSCHENBAUM, H. L. (1996). Treatment regimens for patients with toxoplasmic encephalitis. *Clinical Therapy* **18**, 1037–1056 (discussion 1036).

GARDON, J., GARDON-WENDEL, N., DEMANGA-NGANGUE, KAMGNO, J., CHIPPAUX, J.-P. & BOUSSINESQ, M. (1997). Serious reactions after mass treatment of onchocerciasis with ivermectin in an area endemic for *Loa loa* infection. *Lancet* **350**, 18–22.

GENDREL, D., KOMBILA, M., BEAUDOIN-LEBLEVEC, G. & RICHARD-LENOBLE, D. (1994). Nontyphoidal salmonellal septicemia in Gabonese children infected with *Schistosoma intercalatum*. *Clinical Infectious Diseases* **18**, 103–105.

GOPINATH, R., KEYSTONE, J. S. & KAIN, K. C. (1995). Concurrent falciparum malaria and Salmonella bacteremia in travelers: report of two cases. *Clinical Infectious Diseases* **20**, 706–708.

GRAHAM, S. M., WALSH, A. L., MOLYNEUX, E. M., PHIRI, A. J. & MOLYNEUX, M. E. (2000). Clinical presentation of non-typhoidal Salmonella bacteraemia in Malawian children. *Transactions of the Royal Society of Tropical Medicine and Hygiene* **94**, 310–314.

HEWAN-LOWE, K., FURLONG, B., SIMS, M. & SCHWARTZ, D. A. (1997). Coinfection with *Giardia lamblia* and *Enterocytozoon bieneusi* in a patient with acquired immunodeficiency syndrome and chronic diarrhea. *Archives of Pathology and Laboratory Medicine* **121**, 417–422.

JOHNSON, E. K. & SOAVE, R. (1999) Antiparasitic agents In *Infectious Diseases* (ed. Armstrong DA & Cohen, J) London, Mosby.

KANG, G., MATHEW, M. S., RAJAN, D. P., DANIEL, J. D., MATHAN, M. M., MATHAN, V. I. & MULIYIL, J. P. (1998). Prevalence of intestinal parasites in rural Southern Indians. *Tropical Medicine and International Health* **3**, 70–75.

KATLAMA, C., MOUTHON, B., GOURDON, D., LAPIERRE, D. & ROUSSEAU, F. (1996). Atovaquone as long-term suppressive therapy for toxoplasmic encephalitis in patients with AIDS and multiple drug intolerance. Atovaquone Expanded Access Group. *AIDS* **10**, 1107–1112.

KELLY, P., LUNGU, F., KEANE, E., BAGGALEY, R., KAZEMBE, F., POBEE, J. & FARTHING, M. (1996). Albendazole chemotherapy for treatment of diarrhoea in patients with AIDS in Zambia: a randomised double blind controlled trial. *British Medical Journal* **312**, 1187–1191.

KETTLEHUT, M., EDWARDS, H., MOODY, A. H & CHIODINI, P. L. (1994). The United Kingdom National External Quality Assessment Scheme for Parasitology. *Medical Microbiology Letters* **3**, 203–208.

KOTLER, D. P. & ORENSTEIN, J. M. (1998). Clinical syndromes associated with microsporidiosis. *Advances in Parasitology* **40**, 321–349.

LAMBERTUCCI, J. R., RAYES, A. A. M. & GERSPACHER-LARA, R. (1998). *Salmonella-S. mansoni* association in patients with acquired immunodeficiency syndrome. *Revista do Instituto de Medicina Tropical de Sao Paolo* **40**, 233–235.

LAMBERTUCCI, J. R., RAYES, A. A. M., SERUFO, J. C., GERSPACHER-LARA, R., BRASILEIRO-FILHO, G., TEIXEIRA, R., ANTUNES, C. M. F., GOES, A. M. & COELHO, P. M. Z. (1998). Schistosomiasis and associated infections. *Memorias do Instituto Oswaldo Cruz* **93**, 135–139.

LWAMBO, N. J. S., SIZA, J. E., BROOKER, S., BUNDY, D. A. P. & GUYATT, H. (1999). Patterns of concurrent hookworm infection and schistosomiasis in schoolchildren in Tanzania. *Transactions of the Royal Society of Tropical Medicine and Hygiene* **93**, 497–502.

MARTINELLI, R., PEREIRA, L. J. C., BRITO, E. & ROCHA, H. (1992). Renal involvement in prolonged *Salmonella*

bacteremia: the role of schistosomal glomerulopathy. *Revista do Instituto de Medicina Tropical de Sao Paulo* **34**, 193–198.

MOLINA, J. M., CHASTANG, C., GOGUEL, J., MICHIELS, J. F., SARFATI, C., DESPORTES-LIVAGE, I., HORTON, J., DEROUIN, F. & MODAI, J. (1998). Albendazole for treatment and prophylaxis of microsporidiosis due to *Encephalitozoon intestinalis* in patients with AIDS: a randomized double-blind controlled trial. *Journal of Infectious Diseases* **177**, 1373–1377.

NOKES, C., McGARVEY, S. T., SHIUE, L., WU, G., WU, H., BUNDY, D. A. P. & OLDS, G. R. (1999). Evidence for an improvement in cognitive functioning following treatment of *Schistosoma japonicum* infection in Chinese primary schoolchildren. *American Journal of Tropical Medicine and Hygiene* **60**, 556–565.

PENGGABEAN, M., NORHAYATI, M., OOTHUMAN, P. & FATMAH, M. S. (1998). Efficacy of albendazole in the treatment of *Trichuris trichiura* and *Giardia intestinalis* infection in rural Malay communities. *Medical Journal of Malaysia* **53**, 408–412.

PIERONI, P., MILLS, C. D., OHRT, C., HARRINGTON, M. A. & KAIN, K. C. (1998). Comparison of the ParaSight-F test and the ICT Malaria *Pf* test with the polymerase chain reaction for the detection of *Plasmodium falciparum* malaria in travellers. *Transactions of the Royal Society of Tropical Medicine and Hygiene* **92**, 166–9.

REYNOLDSON, J. A., BEHNKE, J. M., GRACEY, M., HORTON, R. J., SPARGO, R., HOPKINS, R. M., CONSTANTINE, C. C., GILBERT, F., STEAD, C., HOBBS, R. P. & THOMPSON, R. C. A. (1998). Efficacy of albendazole against *Giardia* and hookworm in a remote Aboriginal community in the north of Western Australia. *Acta Tropica* **71**, 27–44.

RICHARD-LENOBLE, D., KOMBILA, M., RUPP, E. A., PAPPAYLIOU, E. S., GOXOTTE, P., NGUIRI, C. & AZIZ, M. A. (1998). Ivermectin in Loiasis and Concomitant *Onchocerca volvulus* and *Mansonella perstans* infections. *American Journal of Tropical Medicine and Hygiene* **39**, 480–483.

SAKTI, H., NOKES, C., HERTANTO, W. S., HENDRATNO, S., HALL, A., BUNDY, D. A. P. & SATOTO. (1999). Evidence for an association between hookworm infection and cognitive function in Indonesian school children. *Tropical Medicine and International Health* **4**, 322–334.

TSHIKUKA, J. G., SCOTT, M. E., GRAY-DONALD, K. & KALUMBA, O. N. (1996). Multiple infection with *Plasmodium* and helminths in communities of low and relatively high socio-economic status. *Annals of Tropical Medicine and Parasitology* **90**, 277–293.

WORLD HEALTH ORGANIZATION (2000). Severe falciparum malaria [Severe and Complicated Malaria, third edition]. *Transactions of the Royal Society of Tropical Medicine and Hygiene* **94**, suppl 1, S1–S90

Printed in the United States
By Bookmasters